The Mystery of Life

This is the exciting story of how life began on earth two and a half billion years ago.

In simple, easy-to-understand language Irving Adler describes the chemical changes and combinations that have helped to channel the direction evolution has taken. He explains what primitive man believed about the origin and nature of life and shows how experiments over the years have changed man's thinking.

He covers current theories resulting from research in such fields as biology, geology, physics, and astronomy, and places special emphasis on chemistry and the nature of carbon.

"One cannot praise too highly the author's narrative skill in presenting his material and his careful definitions as new terms appear. Even formulae lose their terror, so clearly is each explained. His basic accuracy is attested by the fact that Dr. Linus Pauling, Nobel Prize winner in chemistry, writes the preface to the book. Brief and yet profoundly stimulating, this work is a triumph of scientific popularization."

—*Hartford Courant*

IRVING ADLER

HOW
LIFE
BEGAN

Illustrated by RUTH ADLER

With a preface by LINUS PAULING

A SIGNET SCIENCE LIBRARY BOOK
PUBLISHED BY THE NEW AMERICAN LIBRARY

Published as a SIGNET SCIENCE LIBRARY BOOK
By Arrangement with The John Day Company, Inc.

FIRST PRINTING, FEBRUARY, 1959
SECOND PRINTING, JUNE, 1962

How Life Began is published
in England by Dennis Dobson
(Dobson Books, Ltd.), London.

SIGNET TRADEMARK REG. U.S. PAT. OFF. AND FOREIGN COUNTRIES
REGISTERED TRADEMARK—MARCA REGISTRADA
HECHO EN CHICAGO, U.S.A.

SIGNET SCIENCE LIBRARY BOOKS are published by
The New American Library of World Literature, Inc.
501 Madison Avenue, New York 22, New York

PRINTED IN THE UNITED STATES OF AMERICA

Preface

BY DR. LINUS PAULING

*Chairman, Division of Chemistry and
Chemical Engineering, California Institute
of Technology; Winner, Nobel Prize for
Chemistry, 1954*

WHAT IS LIFE? This is an interesting question, a great
question. What is it that distinguishes a living organism,
such as a man, some other animal, a plant, from an inanimate object, such as a piece of limestone? We recognize that
the plant or animal has several attributes that are not pos-
sessed by the rock. The plant or animal has, in general, the
power of reproduction—the power of having progeny which
are sufficiently similar to itself to be described as belonging
to the same species of living organisms. A plant or animal
in general has the ability of ingesting foods, subjecting them
to chemical reactions which usually involve the release of
energy, and secreting some of the products of the reaction.

All of these activities of living organisms are chemical in
nature. To understand them we must understand substances, how they are made up of molecules, how the molecules are made up of atoms.

Closely related to the question of the nature of life is
that of the origin of life. We know that life began on earth
about two billion years ago. We do not yet know, with cer-
tainty, how it began. The problem of the origin of life is
one of the great problems that stimulates man's intellectual
curiosity.

Many scientists have asked about the possible ways in
which atoms could interact with one another so as to give
rise to the first spark of life—a molecule with the power of
reproducing itself.

In the following chapters Irving Adler has discussed the questions of the nature of life and the origin of life in a stimulating way. He has had to talk about atoms and molecules, about elements and compounds, especially about carbon and the compounds in which atoms of carbon are present. He has presented in a simple and yet reliable way the ideas that scientists have had about the nature and origin of life.

I am sure that the young people who have the opportunity to read this book will find it interesting and illuminating, and that by reading it they will greatly increase their understanding of the world in which we live.

Contents

Inside a Cell

The cells of living things are usually too small to be seen with the naked eye. The average cell is so tiny that a line of 500 of them, arranged end to end, would be only one inch long. But small as it is, the

CHAPTER I

The Living and the Dead

A PUPPY BARKS and snaps at a rolling stone as if the stone were another animal. A little child slaps a chair against which he falls, as though he were spanking another child. To animals and young children it seems as though all things are alive. Hundreds of thousands of years ago, when mankind was in its childhood, and was more animal-like in its way of life, it saw the world as animals and young children do. We know this because early man's view of the world is preserved in old stories that have been handed down from generation to generation. In these ancient myths and legends, men spoke of the wind and the trees, of water and fishes, and of fire and birds, as if they were all living things. But after thousands of years of experience men realized that there is an important way in which some things, the animals and plants, differ from other things like rocks, wood, clay pots, or metal knives. Animals and plants are alive, or *animate*. The other things are dead, or *inanimate*. The dead things include plants and animals that used to be alive but aren't any more, and things like rocks and metal, that were never alive at all.

The Breath of Life

After men saw that animals and plants were different from inanimate things, they tried to understand what it was that made them different. They gathered facts about living things, and tried to fit the facts together into theories about what life is and how it began. One of the first theories grew out of a simple fact that men noticed very early. They saw that men breathe as long as they are alive, but their breath leaves them when they die. So they decided that life was the same as breath.

This old belief is hidden in the word *spirit,* which comes from the Latin word meaning to breathe. Primitive man thought of the life or spirit of an animal as a special kind of *thing* that was separate from its body. The body was only a house in which the spirit stayed for a while. Modern science gives us a different view of the nature of life. We know now that breathing plays a part in keeping plants and animals alive. But breath is not life. It is only a flow of air, carrying some gases to the plant or animal, and carrying other gases away. In the view of modern science, life is not a *thing,* separate from the body. It is a *process,* made up of all the movements and activities of the materials within the body. *Biology,* the science of living things, studies this process.

A Potter's View of Life

How did living things come to exist? Man's answer to this question has changed as his knowledge has grown. The first answer, given by primitive man, was based on his own limited experience.

Primitive man, like modern man, was a maker of things. He made arrows, baskets, bowls, huts, boats, and many other useful objects.

Looking at the products of his own labor, he thought he saw a clue to how living things came to exist. Just as bowls and boats existed because he made them, he thought plants and animals existed because they were made by somebody else. He developed the theory that living things were created by gods. His own labor even served as a model for his picture of how it was done. He knew that a potter could mold clay into any shape. He could make not only bowls and jugs, but also little statues in the shape of animals or men. So he thought the gods, too, must have molded animals and men out of clay, and then gave them life by blowing breath into them. We find this belief in the legends of primitive peoples all over the world. A story told by the Bushmen of Australia says that Pund-jel, the creator, made men of clay, and then blew his breath into their mouths, nostrils, and navels. A legend of the Maoris of New Zealand tells how the god Tiki molded man out of clay mixed with blood, and then blew the breath of life into him. Similar stories were told by Eskimos in Alaska, Indians in California, as well as by the ancient Greeks, Egyptians, Babylonians, and Hebrews.

When primitive man made a useful object, he built it according to a design, shaping the parts to fit the purposes for which they would be used. So it was natural for him to think that the gods, too, built according to a design, and made everything to fit a purpose. He combined the theory that living things were made by a creator with the theory that their parts were designed and built to serve a purpose.

Since the days when these early beliefs were formed, man has gathered a tremendous amount of facts about living things. These facts have led scientists to reject the theory that living things were created according to a design. The evidence points to a new type of theory, in which living things are seen as the products of a *process of development*. To explain this process, modern science does not look for a creator who, like man, builds things with a purpose. Instead, science looks for laws of nature that show how causes lead to their effects. In this way science finds out how each step in the process of development leads to the next step.

A Farmer's View of Life

For hundreds of thousands of years man used to get his food by hunting animals and by gathering the fruits, seeds, and roots of wild plants. But while he was engaged in the gathering of food, he made a great discovery. He found that seeds, when planted in the ground, grew up to be plants like those from which he got the seeds in the first place. This discovery made it possible for him to *grow* his food instead of hunting for it. About seven thousand years ago he put his new knowledge to work and became a farmer. At about the

11

same time he also began to breed animals. His work as a farmer or shepherd impressed on him an important fact about living things: plants and animals produce seeds or eggs from which more plants and animals like themselves begin to grow. Life comes from life, and each living thing produces its own kind.

The early farmers were "growing" knowledge as well as crops and livestock. Their work with plants and animals gave them a store of information about how living things grow, how they are fed, and so on. The first scientists of the ancient world used this knowledge as the foundation on which they began to build the science of biology.

Worms, Frogs, Mice and Maggots

The knowledge of the ancient scientists was limited by the weakness of the human eye. They were able to observe things that were large enough to be seen by the naked eye. But where things were too small to be seen by the naked eye, they did not even know that they existed. Their knowledge was also limited by its spotty character. They had bits and snatches of information gathered by accidental observation. They did not carry on thorough, systematic experiments the way scientists do today. Their knowledge was mixed with great amounts of ignorance. The gaps in their knowledge introduced serious errors into their theories.

The ancient scientists knew from the common experience of farmers and shepherds that life produces life, and that each animal and plant produces its own kind. But they also saw what seemed to be evidence that life could be produced by dead matter as well. They saw earthworms coming out of the soil, and frogs emerging from the slime on the surface of a pond. They saw mice suddenly appear, as if from nowhere, in bins where grain was stored. They saw how meat, allowed to stand and rot, was soon covered with white, crawling, wormlike maggots. They did not know that the earthworms grew out of tiny eggs that had been laid by other worms. So they thought the live worms

12

had been produced by the dead soil. They did not know that the frogs grew from jelly-like eggs laid by the parent frogs on the surface of the water. So they thought the frogs had been produced from the slime that they could see. They had not seen how the mice invaded the grain bin in their search for food, so they thought that the grain had produced the mice. They did not know that the maggots grew from tiny eggs laid by flies, and that the maggots themselves would grow up to be flies. So they thought that the maggots had been produced by the rotting meat. What they saw was only part of what really happened. The incomplete picture that they saw led them to the false theory of *spontaneous generation*, the theory that life is always growing out of dead matter, in addition to the life that is produced in the seeds of living things. This theory was accepted by the great Greek scientist and philosopher, Aristotle, who lived about 2,300 years ago. For hundreds of years after his time, the scientists of Europe studied and believed Aristotle's views.

No Eggs, No Maggots

In the year 1668, an Italian physician named Redi covered a piece of meat with some muslin, and then observed it to see if maggots would develop. He saw flies swarming over the muslin. He observed how the flies laid their eggs on the muslin. Then he discovered that maggots grew only if the eggs fell on the meat. As long as the flies were kept from the meat, no maggots developed on it. This was the first of a series of proofs that the theory of spontaneous generation was wrong. The evidence from experiments like Redi's, and growing knowledge about how complicated living things really are, began to convince more and more scientists that living things could not grow suddenly out of dead matter.

Germs in the Broth

While Redi was performing his experiments in Italy, a great discovery was being made in Holland. Leeuwenhoek, using the microscope which he had invented, opened a window on a new world that had never been seen before, the world of things too small to be seen with the naked eye. In this world he found tiny plants and animals. They are called *micro-organisms,* which means very tiny living things. People also sometimes refer to them as *germs.*

Leeuwenhoek turned his microscope on anything he could get his hands on. Among the things he looked at were parts of plants or animals, and things that were made from them. The materials in these things are called *organic,* because they come from living organisms. Everybody knew that organic materials that are allowed to stand begin to spoil. They rot, decay, or ferment. He examined things that had begun to spoil, and he found that the more they decayed, the more crowded they became with micro-organisms. Spoiled beef broth, for example, was soon full of germs, swimming around like fish in an aquarium. This discovery gave fresh support to the theory of spontaneous generation. It seemed to show that decayed or fermented matter was producing living things where none had been before.

The Battle of the Experiments

While some scientists thought that germs were produced by decaying matter, others insisted that the germs really grew from other germs that had fallen in from the air. This disagreement led to a battle of experiments. Some experiments seemed to show that no germs appeared if the organic matter was kept away from air. Others seemed to show that they did appear anyhow. The battle raged for about two hundred years until the question was finally settled by the French scientist, Pasteur. Through a series of careful experiments he proved that there are germs floating in the air, and that they settle on everything around us. He showed that decaying matter did not produce germs. On the contrary, *it was the germs that caused the organic matter to decay.* If the germs were kept away

14

from the organic matter, it would remain unspoiled indefinitely. The studies of earlier investigators had proved that worms, frogs, mice and maggots were not produced by spontaneous generation from dead matter. Pasteur's experiments showed that the same rule applied to micro-organisms. *It is a law of life that living things come only from living things, and like produces like.*

One Big Family

There are hundreds of thousands of different *species* or kinds of animals and plants. A member of any species is produced only by parents that belong to that species. This fact served to emphasize the differences among the species that keep them separate and distinct. At the same time, people could not help noticing that there are resemblances among species, ways in which different species are alike. A wolf and a fox are like a dog in many ways. A tiger, a lion and a leopard are like a cat in many ways. These resemblances suggested that different species may be related. Children in the same family resemble each other because they are descended from the same parents. Cousins often resemble each other because they are descended from the same grandparents. In the same way, when species resemble each other, it suggests that they belong to the same family and are descended from common ancestors. The study of resemblances has shown, in fact, that the wolf, fox, and common dog belong to one *canine* or doglike family, and the tiger, lion, leopard, and cat belong to one *feline* or catlike family. But then there are resemblances among families, too. The dog family and the cat family resemble each other in having hair, "hatching" their eggs inside the mother's body, and producing milk to nurse their young. So their families are related to each other, and belong to a family of families called the *mammals*. Another family of families is made up of all the different kinds of *fishes*. But mammals and fishes resemble each other in that they both have backbones. Resemblances among families of

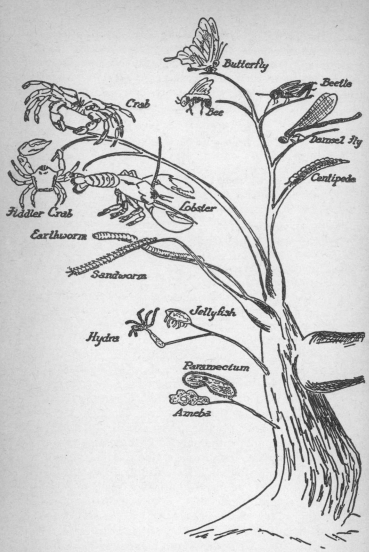

families show that they are related, too. So, mammals are related to fishes. All the resemblances and relationships point to the idea that all living things, whether they are plants or animals, belong to one big family,

16

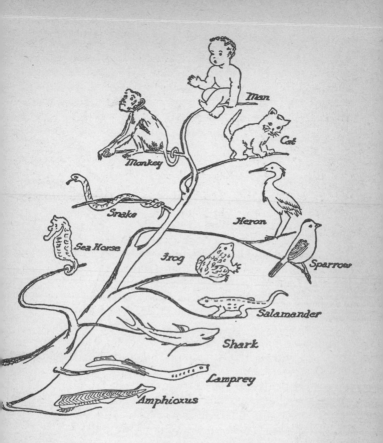

The Tree of Life

and are descended from common ancestors. This is
known as the theory of *evolution*. The family relationships are shown in the Tree of Life drawing on these
pages.

Proof of Evolution

When the theory of evolution was first thought of as an explanation of the family resemblances of plants and animals, it was only a reasonable guess. But by the time it was developed in its present form by the English biologist, Darwin, in 1859, it was no longer just a guess. It was a scientific law proved by many lines of evidence. The first line of evidence was made up of the thousands of resemblances and differences that the Swedish biologist, Linnaeus, used to classify living things into families and families of families. The kinship of all living things was emphasized especially by an important fact that the microscope uncovered. All living things are made up of *cells*. The smallest of the plants and animals consist of a single cell. The larger organisms are made up of many cells. In some, the cells are just neighbors, lying side by side, but each living its own life. In most of the larger plants and animals, the cells work like a team, made up of different groups of cells each of which does a special job to keep the whole community of cells in the body alive. Inside each cell there is a complicated mixture of chemicals called *protoplasm*. Life consists of the activities and processes inside this chemical mixture. The same chief processes take place in the protoplasm of all living things, from the smallest one-celled plants and animals to the largest made up of billions of cells.

Another line of evidence is made up of *fossils*. These are the remains of animals and plants that lived in the past. The bodies of many of them had been covered with mud or sand on the floor of the sea or in river beds. While the soft parts of their bodies decayed, the hard parts like bones and shells sometimes remained. In some cases, the whole body rotted but left its print in the mud or sand. Meanwhile the mud and sand hardened into rock, and in some places the rock was raised out of the water. The rocks that contain fossils are found in layers. The oldest layers are at the bottom. Those formed most recently are near the top. By arranging the fossils that have been found in order, according to the age of the rocks in which they were found, scientists have formed a picture of how life has changed through the ages.

All the complex plants and animals of today are the descendants of simple one-celled organisms that lived over two billion years ago. The first one-celled plants and animals lived in the sea. They grew, and increased their numbers by repeatedly splitting in half and growing some more. They spread all over the world, where they found many different conditions of existence.

Under the influence of these differing conditions they changed in many ways. The offspring of some continued to live as single cells, as we find them today. The offspring of others began to live in clusters and work as a team. Some of these teams stayed in the water, some came out on the land, others took to the air. In the course of millions of years they gradually changed their form. Some old forms, like the *dinosaurs* died out altogether. Some, like the *horseshoe crab,* lived on to the present day, in almost the same form their ancestors had millions of years ago. In the drawing below, *a* is the underside of a present-day horse-

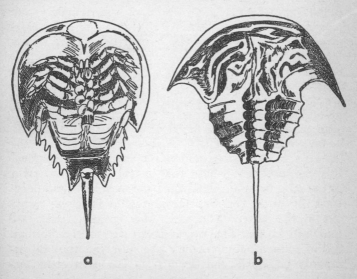

a   b

shoe crab; *b* is a top view of one that was found in a rock 250 million years old. Other forms of life kept changing steadily, so that their present-day descendants are quite different from the forms of life out of which they grew.

A third line of evidence is found in the early life of the more complicated plants and animals. All the larger plants and animals begin life as a single cell. The cell grows and divides many times to form the large number of cells that make up the team in the complete organism. In this first stage of its life the organism is called an *embryo*. In mammals, including human beings, the embryo develops inside its mother's body, until it is born. The embryo of a bird develops inside the egg, until it is hatched. Family relationships among animals or among plants are shown by resemblances in the way in which the embryos develop. The embryo of each species seems to repeat the main steps by which the species developed from the common ancestors of all living things. All mammal embryos, for example, pass through a stage in which they have gills like a fish, showing that mammals are descended from fishlike ancestors.

A fourth line of evidence is found in man's experience as a breeder of animals and cultivator of plants. From the original stock of plants and animals that he domesticated some seven thousand years ago, he has himself developed many new species that differ greatly from their ancestors. The changes that man has made in the forms of living things serve as a small-scale model of the larger changes that nature has produced in two billion years of evolution.

Nature Picks and Chooses

No living thing lives forever. But animals and plants produce offspring like themselves. So although each separate organism dies, its species lives on through its descendants. Yet children, although they resemble their parents, are never exactly like them or like each other. They *vary* in many ways. Among puppies in the same litter, for example, one may have longer ears than the rest. If men who were breeding puppies decided that they liked dogs with long ears, they separated the dogs with the longest ears from the rest, and bred them with each other. By repeating this process through many generations, each time *selecting* for breeding the dogs with the longest ears, they developed dogs with longer and longer ears. The change from one generation to the next was small. But, after many generations, all the small changes added up to a large change. If other breeders took a fancy to long coats of fur, they selected the furriest dogs for breeding, and produced dogs with longer and longer fur. In this way dog fanciers developed many different breeds of dogs from the original stock.

In the same way, nature developed the many species of living things that we see today. Each generation produced offspring like itself, but they varied slightly from each other and their parents. Then nature selected some of the offspring for breeding, and destroyed the others. But nature is not like a breeder who purposely chooses the puppies that have traits that he likes. Nature chooses automatically and without purpose, as a result of *the struggle for existence*. In the world of living things there is severe competition for food, light and air. Animals eat plants, and some animals eat other animals. A plant or animal must survive many dangers in order to live long enough to reproduce. Those who have any serious weakness die young, before they have a chance to have offspring. Their traits which put them at a disadvantage die with them. Those who have some advantage over their competitors or their enemies survive and reproduce, and pass on this advantage to their children. Natural selection weeds out the unfit, and preserves the fit. In this way, each generation becomes slightly more fit for the conditions under which it lives than the generation that preceded it. Slight changes in fitness add up to great changes in the course of millions of years. But what fitness is depends on what conditions the animal is fitted to. Some animals grew up fit for swimming in the sea. Others be-

Life's Time Table

Millions of Years Ago

1,000   500   410   350   280   200

came fit for burrowing in the ground. Still others were fitted for life above the ground, with some fit for crawling, others for running, and some for flying in the air.

Life's Time Table

Life has probably existed on the earth for over two billion years. During the first three quarters of this

170   70

time, it developed from single cells to small, soft, many-celled plants and animals living in the sea. Almost no fossils of this period have been found, either because the soft bodies of the organisms rotted completely, or because the rocks they may be in are buried too deeply for us to find them, or have been worn away by flowing water. But there are many fossils in the rocks that were formed during the last half billion years, and they show us the course that evolution has followed. Five hundred million years ago, living things were already divided into two types, plants and animals. The plants were represented by many algae living in the sea. The animals had developed many forms without backbones, like sponges, worms, jellyfish, and the trilobites, which are relatives of the horseshoe crab. The plants invaded the land about 360 million years ago, in the form of mosses and ferns, which dominated the landscape for over one hundred million years. About 255 million years ago the cone-bearing plants made their first appearance and 165 million years ago the flowering plants came on the scene. Meanwhile the animals, too, produced many different forms. Animals with backbones began to appear 425 million years ago. One hundred million years later, after plants had already invaded the land, animals followed them ashore. Insects, the highest type of animals without backbones, began to appear about 280 million years ago. The first animals with backbones that took to life on land were only part-time land animals. They were the *amphibians* that, like frogs, spent part of their lives in water and part ashore. Some of their descendants became full-time land animals, first in the form of *reptiles*. Amphibians and reptiles overran the land up to about 75 million years ago. By that time they began to give way to mammals and *birds*. Then, about one million years ago, the mammals produced *man*.

The Beginning of Life

The theory of evolution explains how the living things we see today developed from the protoplasm of the first cells. But how did these first cells come into existence? This is the question that is investigated by scientists interested in the origin of life.

About fifty years ago some scientists suggested that the first cells may have developed from life seeds that fell in on the earth from outer space, after being released from some other planet where life exists. Scientists today reject this idea for two reasons. Life probably does not exist on our neighbor planets in the solar system. Therefore no life seeds could have come to us from one of them. If any life seeds were to come to us at all, they would have had to come from a planet that is a satellite of some other star than the sun. To reach us from such another planet, these seeds would have had to travel through space for at least 25 thousand billion miles, which is the distance from the nearest star. But outer space is not empty. It is full of hard-hitting, fast-moving particles that would destroy any life seeds, if they existed, long before they reached the earth. But even if there were such seeds, and they could reach the earth, we would still have to explain how they came into existence on the planet that they came from. To say that they always existed would be dodging the question altogether and giving no answer at all.

The rejection of the life-seed theory leaves only one other possibility. The first living cells must have developed from dead matter, on the earth, some time over two billion years ago. Figuring out how it happened is the greatest detective story of all time. Nobody was there to see it happen. Now, over two billion years after the event, we have to gather what clues we can and try to put them together like a jigsaw puzzle. The clues are drawn from many fields of knowledge. It is necessary to look into astronomy, geology, and the chemistry of the earth to find out what raw materials were at hand on the earth when life began to take shape. It is necessary to look into physics and chemistry to see what processes were at work. It is necessary to look into biology and the chemistry of living things to see how these processes could have produced the protoplasm of living cells. The picture formed from these clues is like a jigsaw puzzle most of whose pieces are missing. We see only parts of the picture, and have a rough idea of how they may be connected. There are many gaps which scientists hope they can fill as they continue their investigations. The remaining chapters of this book describe some of the clues that have already been found, and the picture that they form.

CHAPTER II

Life Activity

Always on the Go

THERE IS CONSTANT movement and activity in living things. It is most obvious in animals, because we see them run, jump, burrow, crawl, swim, or fly. But there is activity even in animals at rest or asleep, and in plants that spend their whole lives in one place. All animals and plants are made up of cells, and the microscope shows that the protoplasm inside a cell is never at rest.

The activity in living things takes many forms. There is *motion,* in which the whole body or part of it moves from one place to another. There is the production of *heat,* most noticeable in mammals and birds, whose bodies are always warm. There is the production of *electrical energy,* as in the small electric currents that flow through nerves, or the large electrical charges built up by the electric eel. There is the production of *light,* as in the flashing body of a firefly. Motion, heat, electricity, and light are forms of *energy.* The living cell is like a factory in which some of these forms of energy are always being produced.

Life Wears Itself Out

No machine can produce energy out of nothing. It can only change energy from one form to another. So it must have a supply of energy to draw from. A gasoline engine, for example, produces energy in the form of motion. In order to produce this motion it must use up energy in another form. The energy it uses is locked up in the form of *chemical energy* in gasoline. By burning the gasoline as fuel, the engine changes the chemical energy to heat, and then changes the heat to motion. But when the gasoline is burned, it is used up. Waste products are formed that are released in the exhaust gases of the engine. A living cell also burns fuel to get the energy that it delivers. Like a gasoline engine, it changes chemical energy into other forms. The fuel that the cell uses is made up of parts of the body of the cell itself. To the extent that fuel is used up, part of the living protoplasm is destroyed, and changed into waste products that are no longer useful to the cell. The cell gets rid of these waste products by releasing them to its surroundings.

When a machine is working, its active parts wear out. In the living cell, too, the parts that are active in carrying out the processes of life are always wearing out. Like the parts of the cell that serve as fuel, they are steadily broken down into waste products that the cell must remove.

A living cell is constantly burning itself up and wearing itself out. Its protoplasm is always being broken down and destroyed. The breaking-down process in the living cell is called *catabolism*.

Life Builds Itself Up

To remain alive, the cell must repair the damage that it does to itself. It must restore its supply of fuel, and rebuild the worn-out parts. So, at the same time that protoplasm burns itself up and wears itself out, it also builds itself up again. The building-up process is called *anabolism*. To carry out this building-up process, the cell must receive from its surroundings a steady supply of fresh materials that it can convert into fuel or working parts. This is the *food* that it needs to keep alive.

The building-up process uses up energy. Like the energy for motion or heat, this energy is obtained from the burning of fuel. But when it is used, it isn't lost to the cell. It is stored in the parts of the cell that are rebuilt.

A Chemical Factory

The breaking-down and building-up processes in protoplasm go on side by side. The combination of

27

these two processes is called *metabolism*. It is the foundation of the activities of life. The changes in the cell that take place during metabolism are largely chemical changes. Complex chemicals in the protoplasm are broken down into smaller and simpler parts. Simple chemicals are reshuffled and combined into more complex parts. The living cell is a chemical factory that is working twenty-four hours a day.

While a factory is in operation, there is a steady flow of materials into and away from the factory. Fuel and raw materials are carried into the factory. Finished products and waste are carried away. While an organism is alive there is also a steady flow of materials into and away from it. Food flows into it to supply fuel and raw materials. But only the waste products are carried away. The finished products that the organism makes are for its own use. It builds them into its own body to replace parts that were worn out before. A living cell is a factory whose product is itself.

Growth and Reproduction

If protoplasm breaks down faster than it is rebuilt, the cell wastes away and dies. But if it is built faster than it breaks down, then the cell grows. There is a limit to how big a cell can grow. When it reaches its greatest size, it splits into two cells. *Growth* in size is changed into *reproduction*, or growth in numbers. Then each of the smaller daughter cells can continue to grow in size. So growth leads to reproduction, and reproduction leads to further growth. In organisms made of more than one cell, growth of the organism is combined with reproduction of its cells. As the individual cells reproduce, the number of cells increases, and the whole organism grows. The organism as a whole is reproduced when one cell, an egg cell, separates from the rest, and, by growing and dividing, develops into a new organism.

Growth and reproduction are special forms of the building-up process within the cells. Since energy is used up in the building-up process, energy is needed for growth and reproduction.

The Materials of Life

The chemical factory inside a living cell produces chemical compounds of a special kind. Because they are made by living organisms, they are called *organic compounds*. We are all familiar with several main types of organic compounds. There are *starches,* like the corn starch we use for cooking and baking. There are *sugars,* like those that give fresh fruits their sweet flavor. There are *fats,* like the butter we spread on our bread, or the vegetable oils in which we fry potatoes. There are *proteins* and *nucleic acids* in the meat we eat.

There are other chemicals found in nature that are not produced by living things. They are called *inorganic* chemicals. Water, table salt, rocks, and the gases in the air, are examples of common inorganic chemicals. Protoplasm is a mixture of both organic and inorganic chemicals. Most of it is water. In the human body, protein makes up about half of the rest.

Overdone Meat

When meat is roasted in the oven, it sizzles as the water in it boils and escapes as steam. If we roast it too long, it loses too much of its water, and becomes dry and tough. If we continue to roast it after it is dry, the meat becomes as black as coal, and we say that it has been burned. The black, coal-like material in the burned meat is *carbon*. It was in the meat in the first place, hidden in the organic compounds of the cells in the meat. The heat broke up these compounds and released the carbon as a fine black powder. *There is carbon in every organic compound.*

We don't have to see meat to know when it is burning. We can *smell* it. There is a very sharp odor when we burn meat, or hair, or any other protein. The sharp odor is caused by compounds of *nitrogen*. The nitrogen was in the meat in the first place. Chemical changes caused by the heat produced the nitrogen compounds that we smell in the smoke of the burning meat. *All protein contains nitrogen.*

The water that was boiled out of the meat is itself a compound. Water contains *hydrogen* and *oxygen*. The four chemical elements, carbon, hydrogen, oxygen, and nitrogen, make up 99 per cent of the material of living things.

Putting Parts Together

Roasting meat is a very crude way of breaking up protein into its parts. Chemists have developed many better ways of doing it. By breaking up organic compounds into their parts, they find out what they are made of. Then they try to find out how the parts are arranged in the complex organic compounds. When they know how the parts are arranged, they may be able to put them together to make or *synthesize* the organic compounds. It used to be thought that organic compounds could be made only by living processes inside living cells. But in the year 1828 the organic compound *urea* was made in the laboratory, from inorganic compounds, by ordinary chemical processes outside of living cells. Since then, many other organic chemicals have been synthesized in the laboratory. The methods for making organic compounds out of inorganic chemicals are important clues to the origin of life, because they show how nature probably produced the first organic chemicals that were built into the bodies of the first living things.

Food Makers

Every living thing consumes food. The food must include organic compounds that it may use as fuel and from which it may build the parts of its body. Most plants get the organic food that they need by *making it themselves* out of inorganic chemicals.

To make organic compounds, plants need a supply

of carbon. They get it from *carbon dioxide,* which they draw out of the air. They combine the carbon dioxide with water to form a sugar called *glucose.* Both the carbon dioxide and the water contain oxygen. Some of this oxygen becomes part of the glucose. The rest is freed and released to the air. By combining the glucose with other chemicals, the plants make all the other organic compounds that they need. When carbon dioxide is combined with water to form glucose, energy is used up and stored in the glucose. So the plant needs a supply of energy in order to be able to make the glucose. It gets the energy it needs from *sunlight.* The process of making glucose with the help of sunlight is called *photosynthesis.* (*Synthesis* means putting together, and *photo* means with light.) Photosynthesis can take place in a cell of a plant only if the cell contains a green chemical called *chlorophyll.* Chlorophyll is responsible for the green color of grass and leaves.

To make the protein it needs, a plant needs a supply of nitrogen. It gets its nitrogen from inorganic compounds of nitrogen that are dissolved in the water that surrounds its roots. It uses the nitrogen to make *amino acids,* and then it combines the amino acids to make proteins.

Plants that make their own organic foods are called *autotrophic,* which means that they *nourish themselves.*

Food Takers

Animals do not have any chlorophyll. But without chlorophyll they cannot make their own organic food. So, to get the organic compounds that they need they feed on the bodies of other animals or plants. They are called *heterotrophic,* which means that they are *nourished by others.* Some animals, like rabbits, are plant eaters. Other animals, like wolves, are meat eaters. Still others, like human beings, eat both plants and meats.

There are plants, too, that do not have any chlorophyll and cannot make their own food. Yeasts, for example, are one-celled plants that feed on the sugar produced by other plants. The feeding action of the yeast is responsible for *fermentation*. Bacteria are one-celled plants some of which feed on the bodies of dead plants or animals. Their feeding action causes the bodies to *decay*. Mushrooms and other fungi are plants that feed on the rotted remains of dead organisms. That is why we find them in the woods on rotting tree trunks, or on ground that has been enriched by rotting leaves.

The First Food Supply

The difference between autotrophic and heterotrophic organisms raises an important question about the origin of life. What kind of organisms were the first living things? Were they food makers or food takers? If they were food makers, then they had to be able to carry out photosynthesis, or some food-making process very much like it. This would have been possible only if their protoplasm contained chlorophyll and other complex chemicals that are necessary for photosynthesis to take place. Photosynthesis, as we shall see in Chapter VI, is a complicated process that takes place in a long chain of steps. It is not likely that the first living things contained such complex chemicals or could carry out such a complicated process.

The first organisms were probably formed out of organic chemicals and processes that were much simpler than those that exist today. The more complicated chemicals and processes must have developed as the result of a long period of evolution after life was already in existence. So it is likely that the first living cells were *not* able to make their own organic food. They were food takers. But, if they were food takers, they needed a ready-made supply of organic compounds to feed on. Since no living cells were making these compounds, there must have been other natural processes that were making organic compounds out of simpler inorganic chemicals. In the next few chapters we shall see that there are such processes that could have taken place when the earth was young, before life began.

Chemical actions in the air and the sea produced the first organic compounds. These organic compounds were the materials out of which the first organisms developed, and served as the first food supply which they consumed. As the organisms increased in numbers and ate more and more, this food supply began to be used up, and food became scarce. The food shortage made the struggle for existence difficult. But, because of this struggle for existence, natural selection and evolution were already at work. Some organisms developed the ability to carry out photosynthesis, and became autotrophic. Their new ability overcame the food shortage. They saved themselves from starvation by *making* the food that they needed. By saving themselves, they also saved the heterotrophic organisms that could live by feeding on them.

Fanning the Fire of Life

The common fire extinguisher puts out a fire by covering it with a blanket of heavy gas. The heavy gas settles right over the fire and prevents air from reaching it. When the fire is deprived of air, it suffocates and dies. This shows that a fire needs air as well as fuel in order to keep burning. Actually it doesn't need all of the air. It needs only the oxygen that is in the air. When the fuel burns, it combines with the oxygen, and, as it does so, it releases the energy that is stored in the fuel. When we fan a fire, we make air flow over the fire. This gives the fire a steady supply of oxygen that keeps it burning brightly.

Animals and plants burn fuel in their cells to get the energy they need. The fire that burns within them is not like an ordinary fire. It is built up out of a series of chemical actions that take place without a flame. But, although the action is different, the result, in most animals and plants, is the same. The fuel is combined with oxygen so that the hidden energy may be released. This can happen only if the organisms have oxygen as well as fuel. That is why animals and plants breathe. By breathing in air, they fan the flame of life that burns within their cells. This process of burning, in which oxygen is used, is called *respiration*.

When coal burns, the carbon in the coal is combined with oxygen to form carbon dioxide. The carbon dioxide escapes from the fire in the smoke. In living cells, when sugar or fats burn as fuel, the same thing happens. The carbon in the fuel combines with oxygen to form carbon dioxide, and the carbon dioxide is released to the air. So, while plants and animals breathe in oxygen, they breathe out carbon dioxide.

Burning without Oxygen

While most animals and plants need oxygen to burn their fuel, there are some that do not. Yeasts, for example, get their energy by fermenting sugar, and fermentation is a process that goes on without drawing oxygen out of the air. Animals and plants that need oxygen from the air are called *aerobic* organisms. Those that do not need air, because they get their energy by fermentation, are called *anaerobic*.

When yeast cells cause sugar to ferment, they turn the sugar into alcohol and carbon dioxide. Because of these products of fermentation, yeast cells are very useful to brewers and bakers. Brewers mix yeast in their mash to produce the alcohol of beer. Bakers put yeast into dough to produce carbon dioxide. The bubbles of carbon dioxide trapped in the dough make it rise.

The difference between aerobic and anaerobic organisms raises a special queston about the origin of life. Which method of burning fuel did the first organisms have? If they were aerobic, then they needed oxygen from the air. But was there oxygen in the air at the time when life began on the earth? We shall see in a later chapter that when the first living cells were formed, there was probably almost no free oxygen in the air. So it is likely that the first living cells were anaerobic. Aerobic organisms could develop only after some process that took place in the past began to put oxygen into the air. We saw on page 31 that photosynthesis is such a process. When green plants

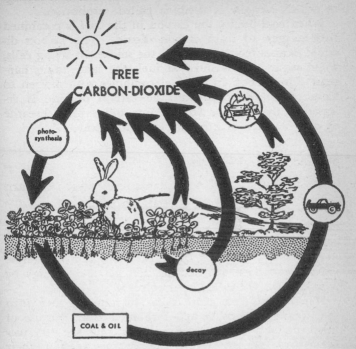

FREE
CARBON-DIOXIDE

photo-
synthesis

decay

COAL & OIL

The Carbon Cycle

make their organic food by photosynthesis, they take
carbon dioxide out of the air, and release oxygen into
the air. This suggests the idea that in the development of early forms of life, the autotrophic organisms,
or food makers that use photosynthesis, had to develop
before any aerobic organisms, the oxygen breathers, could
appear and survive.

The Carbon Cycle

All living things contain carbon in the organic compounds of which they are composed. They get their
supply of carbon from the carbon dioxide in the air.
The green plants remove the carbon dioxide from the

air, and build it into their bodies by photosynthesis. Some of the carbon is built into the bodies of animals when animals eat the green plants. Some of the carbon is hidden in the ground when partly rotted plants are buried and squeezed in swamps to form peat and coal, or when partly rotted animals are buried and squeezed under the sea to form petroleum. So there is a steady flow of carbon, out of the air, into the bodies of plants and animals, and into the underground stores of coal and oil.

But the amount of carbon in the air is very small. Only $3\frac{1}{2}$ per cent of the air today is carbon dioxide. At the rate at which plants are using up carbon dioxide today, the supply would not last very long. However, the supply is never exhausted, because it is being replaced as fast as it is used up. The carbon in organic compounds is being burned to form carbon dioxide again. Some of it is burned as fuel in the living bodies of plants and animals. Some of it is burned in the process of decay, when, after animals and plants die, bacteria and molds make them rot. Some of it is burned when forest fires and fires made by man burn up large amounts of wood, coal, and oil. The carbon dioxide returns to the air, where it is ready to be used by plants once more. In this way the carbon in the air is used over and over again. It travels back and forth between the air and the bodies of living things, like a train going back and forth between the ends of a railroad line.

The Oxygen in the Air

Carbon and oxygen are joined together as partners, in the carbon dioxide of the air. When photosynthesis takes place, the partnership is broken, and the partners are separated. The carbon becomes part of a living plant, and free oxygen is released to the air. When the carbon is burned later on, the free oxygen is rejoined to its partner, and again becomes part of the carbon dioxide. So the carbon cycle is accompanied by an oxygen cycle. The oxygen in the air is joined to carbon, then freed again, and then rejoined over and over again. But the cycle is not completely balanced. There are processes that pull some of the carbon and oxygen out of the cycle and prevent them from joining their old partners.

One of these processes is the formation of coal and oil. When coal and oil were formed hundreds of millions of years ago, the carbon in them was buried in the ground. This prevented the carbon from rejoining the oxygen that used to be its partner. So the oxygen remained in the air in the free state. As more and more carbon was buried, more and more free oxygen accumulated in the air. But not all of the free oxygen remained in the air, because another process pulled some of it out. Oxygen is constantly withdrawn from the air by rocks. In the rocks it combines with iron to form iron rust. So, while more and more oxygen was separated from its old partner, carbon, some of it found a new partner, iron. The rest stayed in the air, and slowly grew in amount. This is how the present supply of oxygen in the air was built up.

The Nitrogen Cycle

All living things contain nitrogen in the proteins which form an important part of their bodies. They get their supply of *nitrogen* from nitrogen compounds called *nitrates* that are found in the soil. Plants draw the nitrates in through their roots, and use them to make *amino acids*. Then they assemble the amino acids to make the proteins they need. Animals get their nitrogen by eating plants or other animals. When they digest their food, they break up the proteins into amino acids, and then rearrange the amino acids to form other proteins of different kinds. So there is a steady flow of nitrogen out of the soil into the bodies of plants and animals. The amount of nitrate in the soil is small, but it is never used up because it is always being replaced. The waste products of animal life and the dead bodies of animals and plants are returned to the soil. There they are attacked by *putrefactive bacteria*

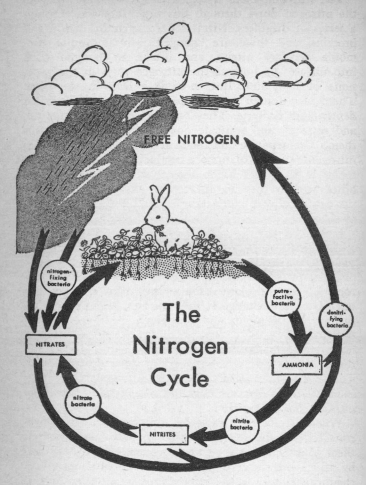

The Nitrogen Cycle

FREE NITROGEN

nitrogen-fixing bacteria

NITRATES

putrefactive bacteria

denitrifying bacteria

AMMONIA

nitrate bacteria

NITRITES

nitrite bacteria

which make them rot. The bacteria change the rotting nitrogen compounds into *ammonia*. Then other bacteria, called *nitrite bacteria,* change the ammonia into compounds called *nitrites*. Then a third type of bacteria, the *nitrate bacteria,* change the nitrites into nitrates. Once the nitrogen is back in the soil in the form of nitrates, it can be used by plants all over again.

The nitrogen cycle is like a closed circular pipe. As the nitrogen flows through this pipe, it passes through a series of disguises. Nitrates become amino acids and proteins, and these are changed into ammonia. Ammonia becomes nitrite, and nitrite is changed back into nitrate. As long as the nitrogen stays in the pipe, none of it is lost. But there is a leak in the pipe. There is a fourth type of bacteria in the soil called *denitrifying bacteria*. They break up ammonia, nitrites, and nitrates, and separate out the nitrogen that is in them. The separated nitrogen is released to the air as nitrogen gas. But plants are unable to use nitrogen gas as food. They can use nitrogen only when it is combined with oxygen in nitrates. So this leak caused by the denitrifying bacteria turns useful nitrogen into useless nitrogen.

If it were unchecked, the nitrogen supply of plants and animals would be drained away. But there are other processes that make up for the loss. A fifth type of bacteria, called *nitrogen-fixing bacteria,* live in the roots of certain plants. There they take nitrogen gas out of the air, change it into nitrates, and put it into the soil. Some molds and blue-green algae are also nitrogen fixers, and change useless nitrogen into useful nitrates. Additional amounts of nitrates are formed in the air by electrical discharges, and are carried down to the ground in rain. The leak in the nitrogen cycle is not stopped but it is overcome, and life's supply of nitrates is maintained.

When farmers cultivate the soil they often draw nitrates out of the soil faster than natural processes put them back. Then it is necessary to make up for the loss by using nitrate fertilizers that are manufactured in chemical factories. In these factories nitrogen gas is combined with hydrogen to form ammonia, and then nitrates are made from the ammonia. The method now used in the factories is one of several methods that nature probably used over two billion years ago to build up the original supply of nitrogen compounds that took part in the formation of the first living things.

Inside a Cell

The cells of living things are usually too small to be seen with the naked eye. The average cell is so tiny that a line of 250 of them, arranged end to end, would be only one inch long. But, small as it is, the cell is a very complicated structure. A typical cell is enclosed by a *cell membrane* which is part of the living proto-

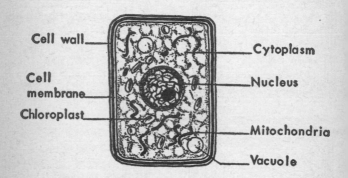

Cell wall — Cytoplasm

Cell membrane — Nucleus

Chloroplast — Mitochondria

— Vacuole

plasm. In plant cells, the membrane is surrounded by a *cell wall* made of dead matter. The protoplasm inside the cell is made up of two main parts, the *nucleus* and the *cytoplasm*. The nucleus is like an island surrounded by a sea of cytoplasm. The cytoplasm is a liquid in which change is going on all the time. Parts of it may be flowing like water, while other parts are stiff and jelly-like. Then the flowing parts may stiffen and the jelly-like parts loosen and begin to flow. There are many things floating in the cytoplasm. There are *mitochondria,* shaped like tiny rods or threads. In plant cells, there are little balls called *chloroplasts,* which contain chlorophyll, the green chemical needed for photosynthesis. The cytoplasm also contains some other structures, as well as granules of many shapes and little droplets of fat. When a cell gets older, *vacuoles,* which are bubbles of clear liquid, begin to form. There are many *membranes,* or skinlike surfaces inside a cell. Some of them, like the cell membrane that surrounds

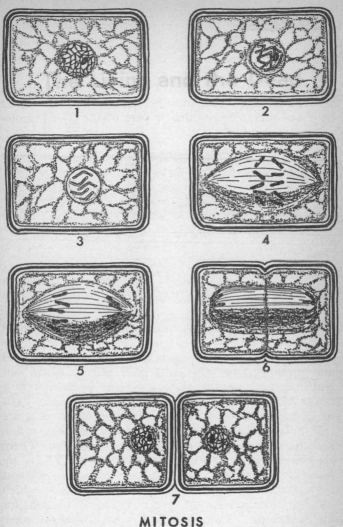

MITOSIS

the whole cell, and the membranes that surround the nucleus and the mitochondria, last a long time. But there are others constantly forming and breaking up in various parts of the cytoplasm.

How a Cell Divides

When a cell divides, it puts on a very interesting show that can be seen under a microscope. The principal actor in the show is the nucleus. The scattered material in the nucleus begins to take shape as a set of distinct threads called *chromosomes*. The chromosomes grow short and fat and arrange themselves near the middle of the cell. Each chromosome splits in half. Meanwhile threads form in the cytoplasm, and begin pulling the halves apart, until the chromosomes are separated into two groups, one at each end of the cell. A membrane develops across the middle of the cell, dividing it into two cells. The chromosomes in each half clump together to form a nucleus. Each of the new nuclei then has as many chromosomes as the original nucleus did. This type of cell division is called *mitosis*. There is another type of division called *meiosis* that takes place when egg cells or sperm cells are formed. In meiosis, the chromosomes do not split before they are separated into two groups. So the egg cell or sperm cell that is formed has only half the original number of chromosomes. In sexual reproduction, an egg cell formed by the mother and a sperm cell formed by the father fuse to form a single cell called the *fertilized egg*. The fertilized egg has the full number of chromosomes again, and a complete organism develops from it as it divides over and over again by mitosis.

Messages in the Cell

The American Indians used to make wampum out of colored beads. The patterns in the wampum were not only decorations. They were symbols with meaning, and could be used to send a message. The chromosomes in the nucleus of a cell are like wampum. They are made up of small particles called *genes,* arranged in line in the chromosomes like beads on a string. The genes contain chemical messages for the living cell. These messages control the metabolism of the cell, and its growth and reproduction. It is as though there were a built-in set of instructions that the cell could read and follow, as building workers read a blueprint when they put up a house. The shape of the fully grown organism and the way that it functions depend on the instructions given by the genes.

When a cell divides by mitosis, the chromosomes split in half. What really happens is that each chromosome builds next to itself another chromosome just like it, and then the two identical chromosomes separate. In this way the instructions built into the original cell are passed on to each new cell that is formed. Egg cells develop into organisms like their parents because, since they have the same kind of chromosomes as their parents, they grow according to the same instructions. Fertilized eggs get half of their chromosomes from the mother and half from the father. That is why children resemble both their parents.

Chemical Regulators

The genes control the development of an organism because they act as chemical regulators. The cell is a chemical factory in which hundreds of chemical changes are taking place. The genes, together with other chemical regulators in the cell, called *enzymes,* control the speed and the order of these chemical changes. The enzymes are like the workmen in the factory, while the genes are like the foremen. The life process in the cell is like production on an assembly line. It is a complicated, delicately balanced, highly organized process in which each step of the process has its proper place and time. The problem to be investigated in the origin of life is not only how the materials of life came into existence, but how they came to be organized into such a complicated and efficient system.

The Simplest Cells

Not all living cells are as complicated as the typical cell described above. Bacteria, for example, do not have a nucleus. They contain nuclear material, but it is scattered in small granules throughout the cytoplasm. When a bacterium divides, there is no complicated process of mitosis. It simply splits in half, with part of the cytoplasm and nuclear material in each half. The blue-green algae also do not have a distinct nucleus. In some of them, the nuclear material is scattered, as in the bacteria. In others, the nuclear material is concentrated in one part of the cytoplasm, but there is no membrane surrounding it. It is likely that the first cells that were formed when life began were like bacteria. The blue-green algae represent steps toward the development of the more advanced cells that have a distinct nucleus.

At the Border of Life

There are certain diseases that are caused by particles that are so small, they pass right through the pores of a porcelain filter. These particles are called *viruses*. Chemically, they are like genes and chromosomes. They also behave like them in that they can reproduce or make copies of themselves when they are inside a cell. To this extent they behave like things that are alive. But when they are not in the cells of an organism that they can infect, they seem to be completely inactive. In fact, they form crystals, just as inorganic chemicals do, and these crystals keep indefinitely. To this extent they behave like things that are not alive. Viruses seem to be in the border area between life and inanimate matter. Some people think that viruses may have been an early stage in the development of living things, and that cells developed when some viruses banded together and surrounded themselves with cytoplasm.

CHAPTER III

Elements and Compounds

Atoms and Molecules

IN THE BRIEF description of life processes in the last chapter, two kinds of chemicals were mentioned. Carbon, nitrogen, hydrogen, and oxygen are examples of chemical *elements*. Water, carbon dioxide, chlorophyll, and ammonia are examples of chemical *compounds*. The elements are the building blocks out of which all chemical substances are made. There are about one hundred different kinds of elements. Compounds are combinations of different elements held together by chemical bonds. There are hundreds of thousands of different compounds, because there are so many different ways in which the elements may be combined.

Any sample of an element or a compound is made up of tiny particles called *molecules*. Each molecule consists of one or more *atoms,* which are the tiniest possible particles of the elements. A water molecule, for example, consists of two atoms of hydrogen joined to one atom of oxygen. This information is shown in its *molecular formula,* a short way of writing how many atoms of each kind there are in the molecule. The formula for water is H_2O. The letter H stands for *hydrogen,* the letter O stands for *oxygen,* and the subscript 2, written after the H, tells us that there are two atoms of hydrogen in the molecule. A molecule of oxygen gas consists of two atoms of oxygen, so it has the molecular formula O_2. Carbon dioxide has the formula CO_2 (one carbon atom and two oxygen atoms in each molecule). Ammonia has the formula NH_3 (one nitrogen atom and three hydrogen atoms in each molecule). The science of *chemistry* studies how compounds are formed, broken up, or changed.

The Shapes of Atoms

The way in which atoms combine to form compounds is explained by the structure of the atoms. Each atom is made up of three kinds of smaller particles, *protons, neutrons,* and *electrons.* The protons and neutrons are crowded together at the center of the atom in the nucleus. The electrons surround the nucleus, and revolve around it the way the earth and the planets revolve around the sun.

The electrons and protons are electrically charged particles. Every electron carries a *negative* charge, and every proton carries a *positive* charge of the same strength. Because of their electrical charges, electrons and protons exert electrical *forces* on each other. Two charges of the same kind tend to *repel,* or push away from each other. Opposite charges tend to *attract,* or pull towards each other. Neutrons are electrically *neutral.* They have no electrical charge, and exert no electrical force.

Because of the protons that are in it, a nucleus has a positive charge. This charge attracts the electrons that surround it, and holds them in place. The more protons there are in a nucleus, the stronger its charge is, and the more electrons it can attract and hold in place. The number of electrons that surround the nucleus of an atom is equal to the number of protons that are in the nucleus. As a result, the negative charge in the atom just balances the positive charge in the atom, and the atom as a whole is electrically neutral. The chemical nature of an element depends on the number of protons in its nucleus, or—what amounts to the same thing—the number of electrons that the nucleus can hold around itself.

Electron Shells

The electrons that surround the nucleus of an atom are arranged in *shells* or layers, like the layers of an onion. The first shell nearest the nucleus has room for two electrons. The second and third shells have room

46

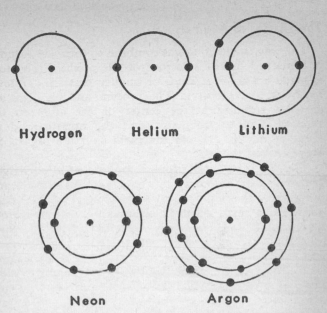

Hydrogen   Helium   Lithium

Neon   Argon

for eight electrons each. If we arrange the atoms of the different elements in order according to the number of protons they have in the nucleus, or according to the number of electrons they have around the nucleus, the first atom would be hydrogen (H), the smallest and lightest of all atoms. It has one proton in its nucleus, and one electron in the first shell. Next in order would be *helium* (He), which has two protons in the nucleus, and two electrons in the first shell. This makes the first shell complete. The next element is *lithium* (Li), which has three protons in the nucleus. It has two electrons in the first shell, and one electron in the second shell. As we continue through the list, each atom has one more electron than the atom just before it, and the second shell is gradually filled up. The second shell is complete in *neon* (Ne), which has two electrons in the first shell, and eight electrons in the second shell. The third shell is begun with *sodium* (Na), which has two electrons in the first shell, eight electrons in the second shell, and one electron in the third shell. The third shell is complete in *argon* (Ar), which has two electrons in the first shell, eight electrons in the second shell, and eight electrons in the third shell. After argon, the structure of the atoms becomes more complicated, because in these larger atoms sometimes electrons are found in an outer shell even though an inner shell is incomplete. The structure of the first eighteen atoms is summarized in the table below.

Element	Symbol	Total Number of Electrons	Number of Electrons in First Shell	Second Shell	Third Shell
Hydrogen	H	1	1		
Helium	He	2	2		
Lithium	Li	3	2	1	
Beryllium	Be	4	2	2	
Boron	B	5	2	3	
Carbon	C	6	2	4	
Nitrogen	N	7	2	5	
Oxygen	O	8	2	6	
Fluorine	F	9	2	7	
Neon	Ne	10	2	8	
Sodium	Na	11	2	8	1
Magnesium	Mg	12	2	8	2
Aluminum	Al	13	2	8	3
Silicon	Si	14	2	8	4
Phosphorus	P	15	2	8	5
Sulphur	S	16	2	8	6
Chlorine	Cl	17	2	8	7
Argon	A	18	2	8	8

Chemical Activity

The chemical behavior of an atom depends largely on the electrons in its outermost shell. If this shell is complete, the arrangement of the electrons is stable, and the atom tends to keep this arrangement without change. If the shell is incomplete, the atom tends to complete it or lose it. An atom with a completed outer shell keeps its electrons undisturbed even when it comes close to other atoms. Atoms of this type do not combine with other atoms to form compounds. They are chemically inactive or *inert*. The inactive elements are the *inert gases*, helium, neon, argon, krypton, and xenon. An atom with an incomplete outer shell completes it or loses it by taking on or giving up electrons. This involves an exchange of electrons with other atoms. So atoms of this type combine with other atoms to form compounds. They are chemically active.

The chemically active atoms whose outer electrons are in the second shell are lithium, beryllium, boron, carbon, nitrogen, oxygen, and fluorine. In the first three of these atoms the outer shell is less than half full. An easy way for these atoms to get a completed outer shell is to lose the electrons in the second shell. Then the first shell, which is already complete, becomes the outer shell. So lithium, beryllium and boron tend to be *electron losers*. Lithium has one electron to lose, beryllium two, and boron three. In the last three atoms in the list, the outer shell is more than half full. These atoms can get a completed outer shell easily by taking on enough extra electrons to give them eight electrons in the second shell. So nitrogen, oxygen, and fluorine are *electron snatchers*. Nitrogen takes on three electrons, oxygen takes on two, and fluorine takes on one, in order to complete the second shell. All atoms that have only a few electrons in their outer shell are electron losers. Those that need only a few extra electrons to complete their outer shell are electron snatchers. Chemists call the electron losers *electropositive elements,* and the electron snatchers *electronegative elements.* Carbon, which is in the middle of the list, can get a completed outer shell by either losing four electrons or gaining four electrons. Its middle position gives it special properties that will be discussed in Chapter IV.

How Atoms Are Held Together

A sodium atom is an electron loser with one electron to lose. A *chlorine* atom is an electron snatcher that needs only one electron to complete its outer shell. So when a sodium atom comes close to a chlorine atom, the sodium atom transfers one electron to the chlorine atom. The sodium atom that has lost an electron is called a sodium *ion*. The chlorine atom that has gained an electron is called a *chloride* ion. Before the transfer of the electron, both the sodium atom and the chlorine atom are electrically neutral, because in each of them the positive charge of the protons in the nucleus is balanced by the negative charge of the electrons around it. But after the electron is transferred, this balance is destroyed. The sodium ion has more protons than electrons, so it has a positive charge. It is called a *positive ion*. The chloride ion has more electrons than protons, so it has a negative charge, and is called a *negative ion*. Since the sodium ion and the chloride ion have opposite charges, they *attract each other*. Each holds on to the other to form a compound. This compound, *sodium chloride,* is ordinary table salt. This is an example of one way in which atoms are held together to form compounds.

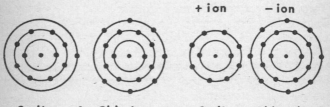

+ ion   − ion

Sodium & Chlorine = Sodium Chloride

When one atom transfers electrons to another, the electron loser becomes a positive ion, and the electron snatcher becomes a negative ion. The oppositely charged ions attract each other. They hold on to each other by a force that is called an *ionic bond*.

Sharing Electrons

It isn't necessary for an atom to lose electrons or take on extra electrons to get a completed outer shell. Atoms can also complete their shells by *sharing* some electrons. For example, a hydrogen atom has one elec-

Hydrogen Atoms

Hydrogen Molecule

Oxygen Atoms

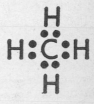
Oxygen Molecule

Water Molecule

Methane Molecule

tron, and needs another to complete the first shell. Two hydrogen atoms can complete their shells by each sharing its electron with the other. Then the same two electrons serve as a completed shell for both atoms. This is how two hydrogen atoms combine to form a molecule of hydrogen gas. When two atoms are held together by sharing electrons, the force that joins them is called a *covalent bond*.

The diagram above shows how two hydrogen atoms share electrons to form the H_2 molecule of hydrogen gas. Each hydrogen atom is represented by the letter H, with a dot next to it to represent the electron in its incomplete outer shell. In the diagram for the hydrogen molecule, the two dots are placed between the H's to show that the two electrons are shared by both atoms.

The diagram also shows how two oxygen atoms share electrons to form the O_2 molecule of oxygen gas. Each oxygen atom is represented by the letter O, with six dots around it to represent the six electrons in its incomplete outer shell. In the diagram for the O_2 molecule, four dots are placed between the O's to show that four electrons are shared by both atoms.

The diagram also shows how two hydrogen atoms and one oxygen atom share electrons to form one molecule of water, and how one carbon atom and four hydrogen atoms share electrons to form one molecule of *methane*. In each case, the dots represent only the electrons of the outer shell.

A shorter way of showing covalent bonds is to represent each pair of shared electrons by a line instead of by two dots, and to leave out the electrons that are not shared. This type of shortened diagram for a molecule is called a *structural formula,* because it shows how the atoms are arranged and joined in the molecule. The molecular formula and structural formula are shown below for hydrogen gas, oxygen gas, water, and methane.

When atoms share electrons, each pair of shared electrons is called a single bond. When two pairs of electrons are shared there is a double bond. When three pairs of electrons are shared there is a triple bond. In hydrogen gas the atoms are held together by a single bond. In oxygen gas the atoms are held together by a double bond.

Part-Time Sharing

There are many compounds in which the electrons that form the chemical bond are neither completely transferred nor completely shared. They spend part of the time near one of the atoms they hold together, and

MOLECULE	HYDROGEN GAS	OXYGEN GAS	WATER	METHANE
MOLECULAR FORMULA	H_2	O_2	H_2O	CH_4
STRUCTURAL FORMULA	H——H	O══O	H-O-H	H-O-H with H above and H below

part of the time near the other. The chemical bond in this case is a compromise between an ionic bond and a covalent bond. It is called a *resonance bond*.

Hydrogen's Little Hook

The electrons that surround the nucleus of an atom serve as a shield. They fence in the positive charge of the nucleus and keep it from extending its influence beyond them into space. But in the hydrogen atom there is only one lonely electron, which finds it hard to "surround" the nucleus all by itself. When a hydrogen atom is joined to a second atom by a covalent bond, its electron is held in place on one side of the nucleus. But then the other side of the nucleus is exposed, and its positive charge can attract negatively charged particles on that side. This gives the hydrogen atom another little hook by means of which it can attach itself to a third atom. Chemists call it the *hydrogen bond*. Hydrogen atoms can form a hydrogen bond only with oxygen, nitrogen, or fluorine. We shall see in Chapter V that the hydrogen bond plays an important part in the structure of protein molecules.

When Electrons Fall

A steam hammer operates by raising a heavy weight and then allowing it to fall. To raise the weight to the higher level from its original position, it is necessary to work against the *force of gravity* which tries to pull the weight down. Energy is used up in doing this work, and it is stored in the position of the weight. When the weight falls to the lower level again, the stored energy is released in the motion of the hammer. In an atom, the positive charge of the nucleus pulls on the surrounding electrons the way the force of gravity pulls on the weight of a steam hammer. When electrons are moved away from the nucleus against this pull, energy is stored in the position of the electrons. The farther the electrons are moved away from the nucleus, the more energy is stored in their position. Then, if the electrons "fall" to a "lower" level, nearer the nucleus, some of this energy is released. When two atoms are about to combine, their electrons start out on a certain energy level. Then, when the atoms combine, the electrons rearrange themselves on a *lower level*. As a result, some of their energy is released, usually in the form of heat. That is why, when atoms unite to form compounds, they give out energy. To separate the atoms again and break up the compound, energy has to be put in to raise the electrons to a higher level again. In more complicated chemical reactions, too, there are some in which energy is released, and others in which energy has to be put in.

Oxidation and Energy

Oxygen is a very active element that unites easily with nearly all other elements to form compounds called *oxides*. The chemical action which makes an oxide is called *oxidation*. When oxidation takes place, energy is released, usually in the form of heat. A fire is simply a place in which something is being oxidized quickly. We make fires in order to get and use the energy that is released. When we burn coal, the carbon in it is oxidized, and we use the energy that is released to heat our homes, cook our food, or run our machines. When we burn gasoline in an automobile engine the carbon and hydrogen in the gasoline are oxidized, and we use the energy that is released to make the automobile go. Living cells burn fuel too, but without a flame. In most cells carbon and hydrogen are oxidized, and the energy released is used for carrying on the business of life.

When carbon is only partly oxidized, each carbon atom combines with one oxygen atom to form *carbon monoxide* CO). When carbon is completely oxidized, each carbon atom combines with two oxygen atoms to form carbon dioxide (CO_2). When hydrogen is oxidized, two hydrogen atoms combine with one oxygen atom to form water (H_2O).

When atoms are oxidized, they transfer some of their

electrons to the atoms of oxygen with which they unite. The same kind of transfer of electrons takes place in many other chemical changes. Because of this resemblance, chemists now use the word oxidation to describe any chemical action in which electrons are transferred, even if the atoms that gain the electrons are not atoms of oxygen. They say that the atoms or molecules that lose electrons are oxidized, and those that gain electrons are *reduced*. To emphasize the fact that when an electron transfer takes place one atom's gain is another atom's loss, they call the transfer an *oxidation-reduction reaction*.

Water

One of the most important of the oxides we know is water. We drink it when we are thirsty. We wash ourselves with it when we are dirty. Most of it is in the sea, where it covers seven tenths of the surface of the earth. From the sea some of it rises into the air as vapor, and then falls again as rain.

Water is especially important for life. Life began in water, probably over two billion years ago. As a result, there are living things in water, and there is water in all living things. Protoplasm is a complicated mixture of chemicals dissolved or floating around in water. Many of the chemicals themselves are built out of water, because water and carbon dioxide are the chief raw materials out of which organic chemicals are made.

A water molecule consists of two hydrogen atoms and one oxygen atom. In most of the molecules of liquid water, the hydrogen atoms *share* their electrons with the oxygen atom, and the molecule is held together by covalent bonds. But in some of the molecules one of the hydrogen atoms *transfers* its electron to the rest of the molecule. When this happens the hydrogen atom that loses its electron becomes a positive hydrogen ion. The rest of the molecule, made up of one hydrogen atom joined to an oxygen atom, is known as the *hydroxyl group*, OH. When this group gains the electron, it becomes a negative hydroxyl ion.

55

The Dance of the Molecules

The molecules in a liquid are always moving. They dart this way and that, like dancing couples on a dance floor. The liquid is crowded, so the molecules often collide with their neighbors. As a result, the ions are pushed around, and are sometimes separated from their partners. Now and then a hydrogen ion may be pushed close to a hydroxyl ion, and will rejoin it to form a molecule of water. The number of ions that remain unattached is always small compared to the total number of molecules. In fact, if, in some way, a large number of hydrogen ions and a large number of hydroxyl ions are brought together like men and women at a dance, most of them join up with partners to form water molecules, and only a small number remain alone, surrounded by their dancing neighbors.

The dance of the molecules in liquid water is made more complicated by the hydrogen bond. The hydrogen bond is like a little hook that makes it possible for hydrogen that is already part of a molecule of water to attach itself to the oxygen in another molecule. Through the hydrogen bond, the molecules of water join together in bunches, and bunches join to bunches. Some molecules may join a bunch while others break away. The liquid water is like a jam-packed dance floor where a pushing, pulling, milling throng are doing the lindy, a square dance, and the conga all at once.

Changing Partners

In some water molecules, a positive hydrogen ion is joined to a negative hydroxyl ion as its partner. If the hydroxyl ion is replaced by a different kind of negative ion, a new compound is formed in which the positive hydrogen ion has another partner. Such a compound is called an *acid*. The negative ion in the acid may be an element that has gained electrons, or, like the hydroxyl ion, it may be a group of elements that has gained some electrons. For example, a chlorine atom that has gained one electron becomes the chloride ion.

When joined to a hydrogen ion, it forms *hydrochloric acid*. A group made up of one nitrogen atom joined to three oxygen atoms (NO_3) and with one extra electron forms the nitrate ion. When the nitrate ion is joined to a hydrogen ion, it forms nitric acid. The nitrite ion (NO_2) joined to a hydrogen ion forms *nitrous acid*. The sulphate ion (SO^4), joined to two hydrogen ions, forms *sulphuric acid*. The carbonate ion (CO_3), joined to two hydrogen ions, forms carbonic acid. The *phosphate* ion (PO_4), joined to three hydrogen ions, forms *phosphoric acid*. The phosphate ion is sometimes combined with one or two hydrogen ions to form other negative ions with a weaker charge.

If the hydrogen ion in a water molecule is replaced by a different kind of positive ion, a new compound is formed in which the negative hydroxyl ion has another partner. Such a compound is called a *base*. The positive ion in the base may be an element that has lost electrons, or it may be a group of elements that has lost some electrons. For example, the positive sodium ion, joined to the negative hydroxyl ion, forms the base called *sodium hydroxide*. An atom of *ammonia* (NH_3) can join a hydrogen ion to form the positive ammonium ion (NH_4). An ammonium ion, joined to the hydroxyl ion, forms the base *ammonium hydroxide*, usually known as ammonia water and used for cleaning purposes.

If a positive ion other than the hydrogen ion is joined to a negative ion other than the hydroxyl ion, the compound formed is called a *salt*. The best known salt is the table salt we use to flavor our food. It is made of sodium ions joined to chloride ions.

Invisible Salt

If you shake some crystals of table salt into a glass of water and stir the water, the salt crystals disappear. We say that the salt is *dissolved* by the water. What happens is that the molecules of salt in the crystals are caught up in the dance of the molecules of water, and then are separated and scattered. Some of the molecules are even broken up. Their ions are separated and wander around without partners until, by chance, a positive ion is thrown close enough to a negative ion to form a partnership again. We can see a crystal of salt because it has millions of molecules crowded together. We cannot see the salt that is dissolved in water because the molecules are scattered, and each molecule or ion is too small to be seen by itself.

There are many compounds that can dissolve in water the way table salt does. In fact, water dissolves so many things, it has been called the universal solvent. Rain water dissolves gases in the air, and carries them down to the ground. On the ground it dissolves compounds that are in the soil or the rocks, and carries them down to the sea. The sea is like a giant test tube in nature's chemical laboratory, where chemicals brought from all corners of the earth are dissolved and mixed. *Life developed out of the chemical mixtures that were in the sea over two billion years ago.*

When Acid Meets Base

If hydrochloric acid is dissolved in water, its molecules are scattered, and some of them are broken up into ions. So, in the solution, hydrogen ions and chloride ions exist side by side with hydrochloric acid molecules. Some of the ions may rejoin to form molecules, but they are constantly replaced by ions from other molecules that are broken up. If the base, sodium hydroxide, is added to the solution, the sodium hydroxide molecules scatter, and some of them are broken up into sodium ions and hydroxyl ions. Then the solution contains two kinds of positive ions, sodium and hydrogen, and two kinds of negative ions, chloride and hydroxyl. Each of the positive ions may join with each of the negative ions to form a molecule. But when the hydrogen ions join with hydroxyl ions, water molecules are formed, and very few of these break up again into ions. As more molecules of the acid and the base break up, more molecules of water are formed. If there were originally equal numbers of molecules of the acid and base, they all finally break up and their hydrogen and hydroxyl ions unite and remain united as water. The chloride ions and the sodium ions, having nothing else to unite with, join with each other to form table salt. So what started as a solution of an acid and a base ends as a solution of salt. *This happens whenever an acid and a base are dissolved together. The hydrogen ion of the acid and the hydroxyl ion of the base combine to form water, and the other two ions combine to form a salt.*

Carbides

When carbon behaves as an electron snatcher, carbon atoms can combine with the atoms of metals to form compounds called *carbides*. When the earth was young, melted iron, nickel and cobalt formed, and sank to the center of the earth. Carbon that was mixed with the metals united to form carbides, so there are carbides deep in the interior of the earth.

When carbide molecules are in contact with water molecules, they interact, and the atoms in the molecules change partners. The carbon separates from the metal that it is joined to in the carbide, and combines with hydrogen that it gets from the water. The result is a compound of hydrogen and carbon called a *hydrocarbon.* For example, when *calcium carbide* interacts with water, the carbon joins with hydrogen to form the hydrocarbon *acetylene,* whose molecular formula is C_2H_2. Meanwhile the calcium combines with the hydroxyl group of water to form the base *calcium hydroxide.* Acetylene is a gas that can be burned in a gas stove.

When the earth was young, material from the interior of the earth often broke through the crust of the earth in volcanic eruptions. Carbides brought to the surface in this way must have interacted with water in the air to form hydrocarbons. When the first rains began to fill the ocean basins, they carried these hydrocarbons into the sea. Hydrocarbons, as we shall see in the next chapter, are simple organic compounds out of which more complicated organic compounds can be built. *When hydrocarbons were first brought into the sea, a big step was taken toward the development of the organic compounds out of which living things were formed.*

Carbides mixed with nitrogen gas at a high temperature will interact with the nitrogen to form nitrogen compounds called *cyanamides.* For example, when calcium carbide reacts with nitrogen, some of the carbon in the carbide is freed, but the rest joins with the calcium and the nitrogen to form the fertilizer calcium cyanamide. *When the earth was young, carbides brought to the surface by volcanic eruptions could have combined with the nitrogen in the air. In this way cyanamides were formed at the surface of the earth.*

Ammonia Makers

Cyanamides are important as ammonia makers. For example, when calcium cyanamide interacts with water, the calcium, carbon and oxygen combine to form *calcium carbonate,* while the nitrogen and hydrogen combine to form ammonia. This could have happened with the cyanamides formed at the surface of the earth billions of years ago. After the carbides that came to the surface from inside the earth united with nitrogen to form cyanamides, the cyanamides interacted with water to form ammonia. *The early appearance of ammonia was important for the origin of life, because ammonia was needed for the formation of amino acids, and amino acids are needed to make proteins.*

There are other processes, too, that are now used in laboratories and factories for making ammonia. Some of these processes could have taken place billions of years ago to add to the supply of ammonia on the earth. Just as carbon combined with metals to form carbides, nitrogen also combined with some metals like lithium, calcium and magnesium, to form *nitrides.* Then the nitrides, brought to the surface, interacted with water to form ammonia. Meanwhile, high in the atmosphere, ammonia could also have been formed directly from the nitrogen gas and hydrogen gas that were there.

Speeding Up a Change

Ordinary cane sugar burns by combining with oxygen. This happens on the surface of a lump of sugar, where the sugar is in contact with the air. But it happens so very slowly that practically none of the sugar is oxidized. Even if you hold a lump of sugar in the flame of a match, it burns so slowly that it does not seem to burn at all. But if you put some cigarette ashes on the sugar before you hold it in the flame, the sugar catches fire and burns with a flame. The tiny particles of ashes have the effect of speeding up the process of burning the sugar. The ashes are an example of what chemists call *catalysts,* substances that can be used to increase the speed of a chemical reaction.

The process in which nitrogen gas and hydrogen gas are combined to form ammonia is important commercially because nitrogen and hydrogen are cheap and easy to get. But the process is practical only if it takes place fast enough to produce a lot of ammonia. To speed up the process, finely divided iron is used as a catalyst. The iron becomes an even better catalyst if it is mixed with tiny amounts of the oxides of sodium and aluminum. These oxides are called *promoters,* because they speed up the action of the catalyst. Catalysts and their promoters must be kept clean, because there are other substances called *inhibitors,* which can hold back the action of the catalysts.

The chemical changes in living things take place at high speed, because they too are regulated by catalysts and promoters. The catalysts in living organisms are the enzymes that were mentioned in Chapter II. The enzymes are special proteins. Each has its own special job to do, and they all co-operate to regulate the order and speed of the chemical changes that take place in living cells. *Part of the job of explaining the origin of life will be to explain how the enzymes that life depends on came to exist.*

CHAPTER IV

Carbon and Its Family

The Sociable Atom

ALL ORGANIC COMPOUNDS contain carbon. In fact, organic chemistry is simply the chemistry of carbon compounds.

A carbon atom has six electrons surrounding its nucleus. Two of these electrons are in the first shell nearest the nucleus. The other four electrons are in the second shell. But the second shell has room for eight electrons altogether. So, in carbon, the second shell is only half full. This fact gives carbon a very special place among the elements. In atoms of lithium, beryllium and boron, the second shell is less than half full, so they tend to combine with other atoms by losing electrons. In atoms of nitrogen, oxygen, and fluorine, the second shell is more than half full, so they tend to combine with other atoms by gaining electrons. The carbon atom, because of its middle position, usually tends to combine with other atoms by sharing electrons. In addition, while the other elements may gain or lose one, two, or three electrons, carbon has four electrons to share. With each of these electrons it can form a covalent bond, and so one carbon atom can form bonds with as many as four other atoms at once.

Carbon atoms can also join with other carbon atoms to form a chain. Since each carbon atom uses up only two bonds to connect it with the two carbon atoms that are on either side of it in the chain, it still has two bonds left for hooking on to other atoms. For all these reasons, carbon is the most sociable of the atoms. It can form many more compounds than other elements can. For example, in combining with hydrogen atoms alone, lithium, beryllium, and fluorine can form only one compound each; oxygen can form 2; boron and nitrogen can form 7; but carbon can form over 2300. Carbon chains can also combine with many atoms at a time to form large molecules, and the compounds can interact in a large number of ways, exchanging atoms and energy.

This sociability of the carbon atom is the basis of the origin of life. The story of how life began is the story of how carbon compounds became more and more complicated and entered into more and more complex processes, until finally systems of interacting carbon compounds developed that had the property of being alive.

Chains and Rings

There are some compounds in which carbon atoms are joined to each other to form a *straight chain*. Other atoms are attached to the chain along the sides and at the ends. For example, a molecule of *octane,* one of the compounds in gasoline, is a chain of eight carbon atoms, with eighteen hydrogen atoms attached. The structural formula below shows how each carbon atom is connected by its four covalent bonds to four neighboring atoms.

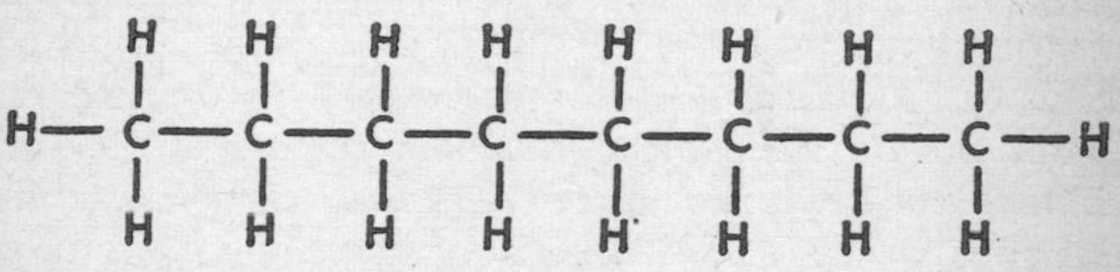

OCTANE

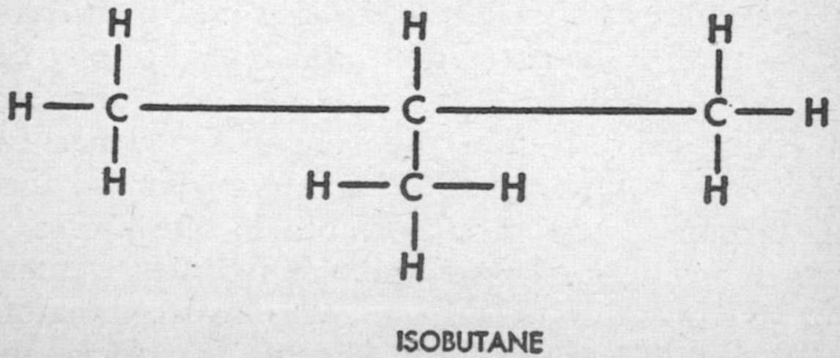

ISOBUTANE

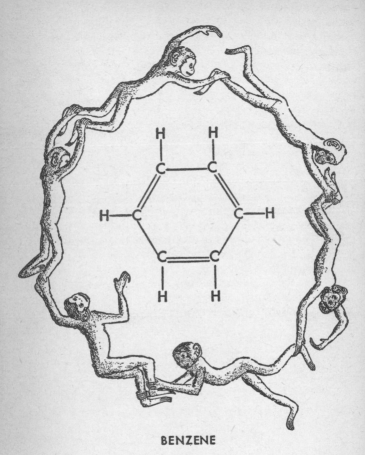

BENZENE

If a carbon atom in a chain is connected to *three* other carbon atoms instead of only two, then the molecule is a *branched chain*. In *isobutane,* for example, there is a chain of three carbon atoms, with another chain branching off from the carbon atom in the middle.

In some compounds, the ends of the chain are joined to form a closed *ring*. In *benzene,* for example, there are six carbon atoms in a ring, and each is attached to a hydrogen atom. Each carbon atom is joined by a single bond to one carbon atom and by a double bond to another. A German chemistry magazine published in 1886 showed this structure in a humorous way by representing the carbon atoms by monkeys. Where the carbon atoms are joined by a single bond, a monkey uses one hand to hold one leg of the monkey ahead of him. Where the atoms are joined by two bonds, he uses both hands to grasp both legs. The hands and legs that are free represent the electrons that form bonds with the six hydrogen atoms.

While in the benzene molecule only carbon atoms are in the ring, there are some rings that contain other atoms as well. The *pyrrole* ring, for example, contains one nitrogen atom. *Pyrimidine* has two nitrogen atoms in the ring. More complicated molecules are formed when one ring is attached to another so that some atoms belong to more than one ring, as in *purine*.

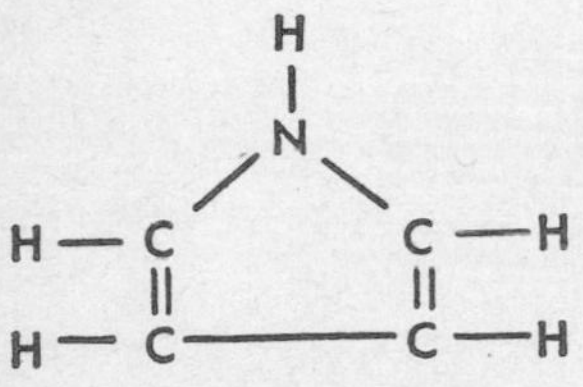

PYRROLE

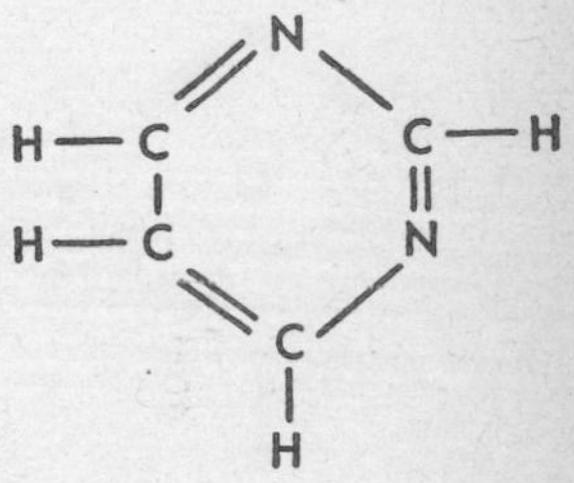

PYRIMIDINE

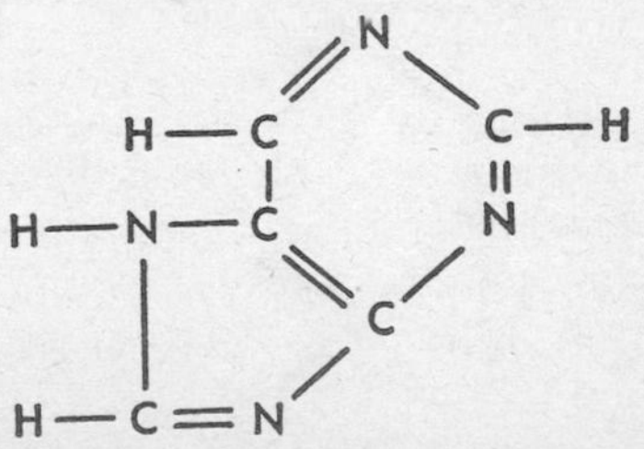

PURINE

Carbon and Hydrogen

Compounds that contain only carbon and hydrogen atoms are called hydrocarbons. Acetylene, formed when calcium carbide interacts with water, is an example of a hydrocarbon. Its molecular formula is C_2H_2. The structural formula for acetylene shows that its carbon atoms are joined to each other by a triple bond, and each of the carbon atoms is attached to only two other atoms. If acetylene is heated with hydrogen, with finely divided nickel serving as a catalyst, one of the bonds between the carbon atoms is broken, and then each carbon atom uses the broken bond to attach itself to another hydrogen atom. The result is *ethylene,* whose molecular formula is C_2H_4. In ethylene, the carbon atoms are joined to each other by a double bond, and each carbon atom is attached to three other atoms. If

ACETYLENE $\quad$ H — C ≡ C — H

ETHYLENE

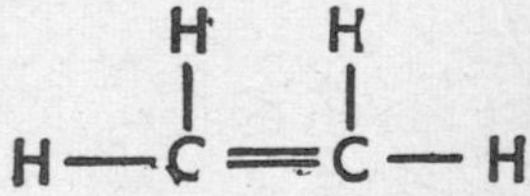

$$H-\overset{\displaystyle H}{\underset{}{C}} = \overset{\displaystyle H}{\underset{}{C}}-H$$

ETHANE

$$H-\overset{\displaystyle H}{\underset{\displaystyle H}{C}} - \overset{\displaystyle H}{\underset{\displaystyle H}{C}}-H$$

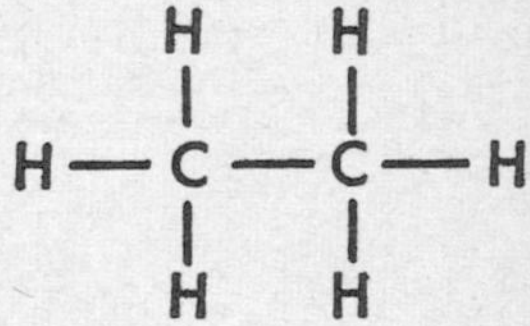

ethylene is heated with hydrogen, in the presence of finely divided nickel, another bond between the carbon atoms is broken, and again each carbon atom uses the broken bond to attach itself to another hydrogen atom. The result is *ethane,* C_2H_6, in which the two carbon atoms are joined by only a single bond, and each carbon atom is attached to four other atoms. In the change from acetylene to ethylene to ethane, the number of atoms that the carbon atoms were attached to was increased. It cannot be increased any more, because a carbon atom can form only four bonds, and so can be attached to at most four other atoms. Ethane, in which the carbon atoms form only single bonds, and are joined to as many atoms as they can hold, is called a *saturated* compound. Ethylene and acetylene, in which the carbon atoms form double or triple bonds with each other are called *unsaturated* hydrocarbons, because the carbon atoms in them are attached to less than four other atoms each.

The Paraffin Series

The saturated hydrocarbon, ethane, is a straight chain compound that belongs to a family known as the *paraffin series*, whose members are found mixed to-

METHANE

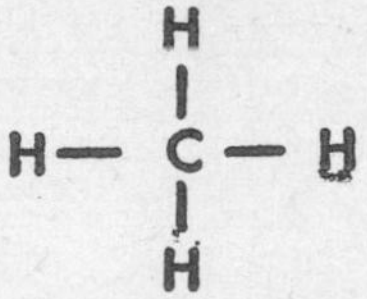

$$H - \overset{\displaystyle H}{\underset{\displaystyle H}{\overset{|}{\underset{|}{C}}}} - H$$

ETHANE

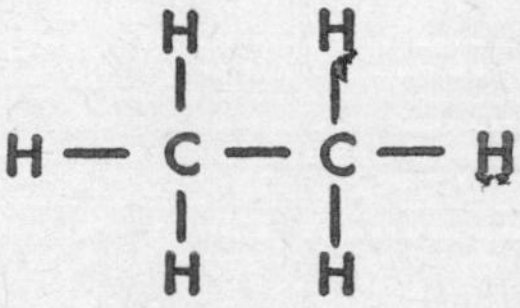

$$H - \overset{\displaystyle H}{\underset{\displaystyle H}{\overset{|}{\underset{|}{C}}}} - \overset{\displaystyle H}{\underset{\displaystyle H}{\overset{|}{\underset{|}{C}}}} - H$$

PROPANE

$$H - \overset{\displaystyle H}{\underset{\displaystyle H}{\overset{|}{\underset{|}{C}}}} - \overset{\displaystyle H}{\underset{\displaystyle H}{\overset{|}{\underset{|}{C}}}} - \overset{\displaystyle H}{\underset{\displaystyle H}{\overset{|}{\underset{|}{C}}}} - H$$

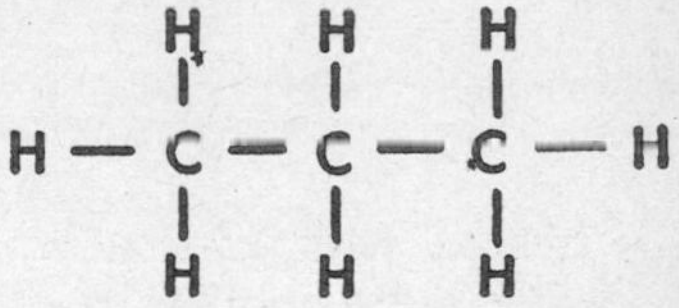

BUTANE

$$H - \overset{\displaystyle H}{\underset{\displaystyle H}{\overset{|}{\underset{|}{C}}}} - \overset{\displaystyle H}{\underset{\displaystyle H}{\overset{|}{\underset{|}{C}}}} - \overset{\displaystyle H}{\underset{\displaystyle H}{\overset{|}{\underset{|}{C}}}} - \overset{\displaystyle H}{\underset{\displaystyle H}{\overset{|}{\underset{|}{C}}}} - H$$

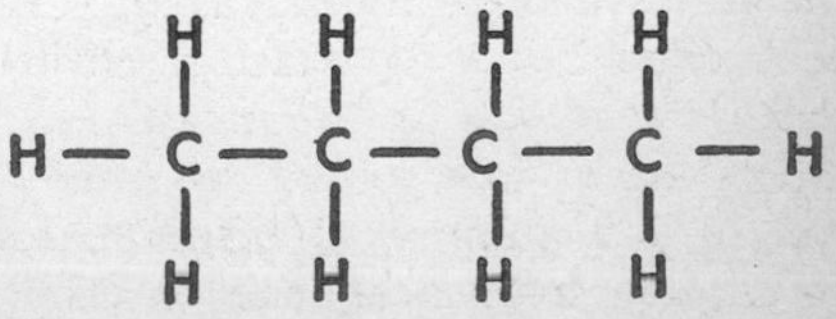

gether in petroleum. In ethane, there are two carbon atoms in the chain. The chain can be shorter, with only one carbon atom in it, as in *methane*, or longer, as in *octane*, which has eight carbon atoms in the chain. If the members of the series are arranged in order according to the length of their carbon chains, the first four are methane, ethane, *propane* and *butane*. Their structural formulas are shown on page 67. The atoms in butane can also be rearranged to form a branched chain, as in isobutane, whose structural formula appears on page 63.

Pulling Away Electrons

The hydrocarbons are the simplest organic compounds. Other compounds can be formed from them by removing hydrogen atoms from the molecules and putting other atoms or groups of atoms in to take their place. If, for example, a hydrogen atom is replaced by a hydroxyl group (OH), the resulting compound is called an *alcohol*. When OH replaces one hydrogen atom in methane, we get *methyl alcohol*. When OH replaces one hydrogen atom in ethane, we get *ethyl alcohol*.

METHYL ALCOHOL

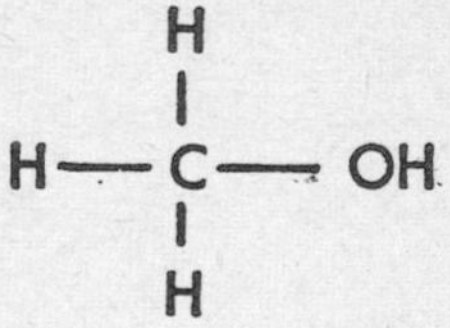

ETHYL ALCOHOL

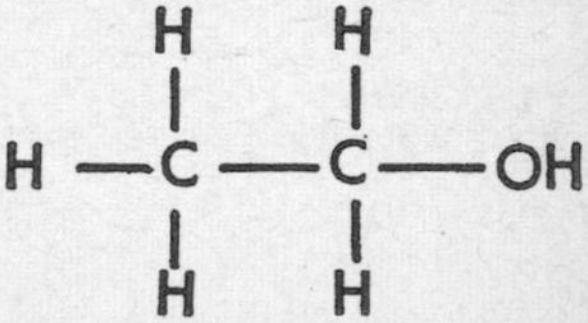

The bond between a hydrogen atom and a carbon atom in a hydrocarbon is formed by two electrons that they share. When OH replaces hydrogen in the molecule, it pulls this pair of electrons further away from

68

the carbon atom than it was before. But when electrons are pulled away from an atom, the atom is oxidized. So, when an alcohol is formed from a hydrocarbon, the carbon in it is oxidized.

A carbon atom that forms four bonds has four pairs of electrons that can be pulled away from it to a greater distance. If only one pair is pulled away, the carbon atom is in the *first stage* of oxidation. If two pairs are pulled away, the carbon atom is in the *second stage* of oxidation. If three pairs are pulled away, the carbon atom is in the *third stage* of oxidation. If all four pairs are pulled away, the carbon atom is in the *fourth and final stage* of oxidation. Alcohols are compounds in which carbon is in the first stage of oxidation.

Some Alcohols

When alcohols are formed from long chain hydrocarbons, it makes a difference which carbon atom the hydroxyl group is attached to. If it is attached to a carbon atom at the end of the chain, the result is called a *primary alcohol*. If it is attached to a carbon atom that is in the middle of the chain, the result is called a *secondary alcohol*. The structural formulas for *primary butyl alcohol* and *secondary butyl alcohol*, both derived from butane, are shown below.

PRIMARY BUTYL ALCOHOL

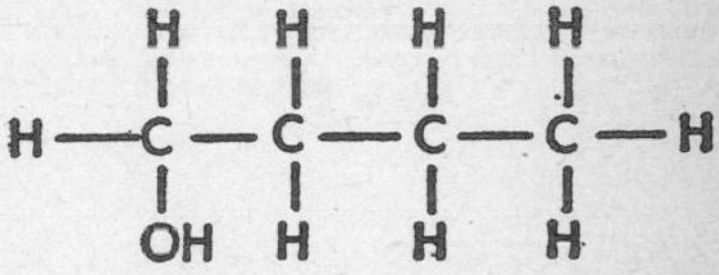

SECONDARY BUTYL ALCOHOL

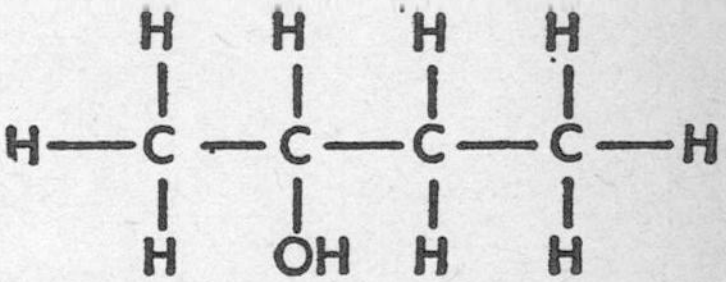

GLYCEROL

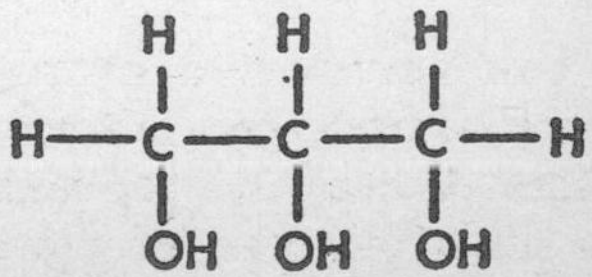

There are also alcohols in which more than one carbon atom has an OH group attached to it. *Glycerol,* for example, has three carbon atoms, just as propane has. Each of them has an OH group in place of a hydrogen atom, so that each of the three carbon atoms is in the first stage of oxidation.

Substitutes for OH

There are other groups of atoms that can take the place of the OH group, when the carbon is in the first stage of oxidation. One of these is the amino group, NH_2. When a hydroxyl group is replaced by an amino group, the result is called an *amine*. Amines can be formed from the reaction of alcohols with ammonia.

Second Stage

In a primary alcohol, the carbon atom that is joined to the OH group is also joined to two hydrogen atoms. If the OH group and one of these hydrogen atoms are removed, the carbon atom is left with two broken bonds with which it can hook on to something else. If both bonds are joined to an oxygen atom, the result is called an *aldehyde*. In an aldehyde, two pairs of electrons are pulled away from the carbon atom, so it is in the second stage of oxidation. Aldehydes can be obtained from alcohols by treating them with an oxidizing agent. When ethyl alcohol is oxidized, *acetaldehyde* is formed. Its structural formula is shown below.

ACETALDEHYDE

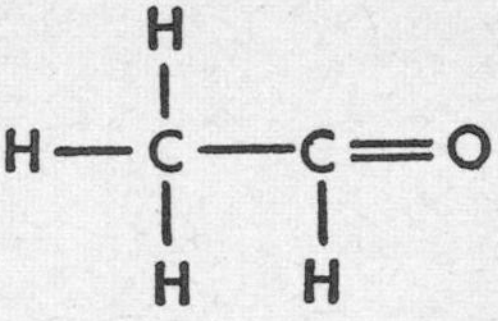

$$H-\overset{\overset{\displaystyle H}{|}}{\underset{\underset{\displaystyle H}{|}}{C}}-\overset{\overset{\displaystyle H}{|}}{C}=O$$

ACETONE

$$H-\overset{\overset{\displaystyle H}{|}}{\underset{\underset{\displaystyle H}{|}}{C}}-\overset{\overset{\displaystyle H}{|}}{\underset{\underset{\displaystyle O}{|}}{C}}-\overset{\overset{\displaystyle H}{|}}{\underset{\underset{\displaystyle H}{|}}{C}}-H$$

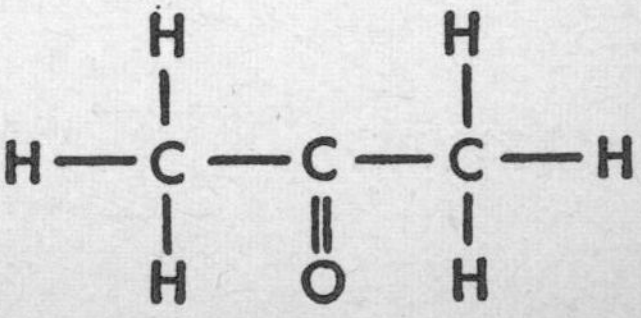

In a secondary alcohol, the carbon atom that is joined to the OH group is also joined to one hydrogen atom. If the bonds to the OH group and the hydrogen atom are broken and then attached to an oxygen atom, the result is called a *ketone*. In a ketone, as in an aldehyde, the carbon atom that is joined to the oxygen atom is in the second stage of oxidation. Ketones can be formed by oxidizing secondary alcohols. The structural formula for *acetone*, a three-carbon ketone, is shown on page 70.

In both aldehydes and ketones, a carbon atom is joined by a double bond to an oxygen atom. This linked pair of atoms is called the *carbonyl group* (CO).

Fatty Acids

In an aldehyde, a carbon atom at the end of a chain is joined to an oxygen atom and to a hydrogen atom. If the hydrogen atom is replaced by the hydroxyl group, OH, then the carbon is linked to O and OH, forming the *carboxyl group*, COOH. A molecule that contains the carboxyl group is called a *carboxylic acid*. The carbon in the carboxyl group is in the third stage of oxidation. Carboxylic acids derived from the saturated compounds of the paraffin series are called *saturated acids* or *fatty acids*. There are also *unsaturated*

ACETIC ACID

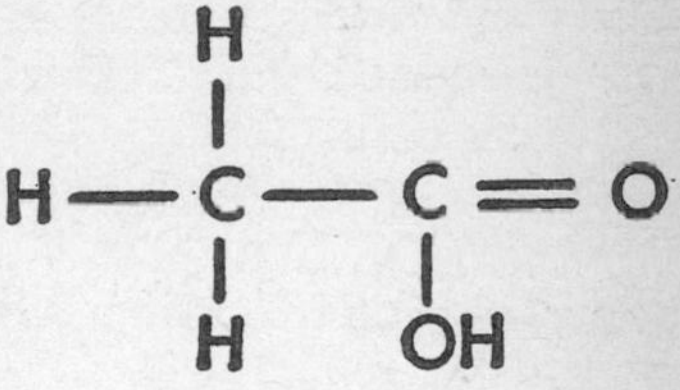

$$H - \underset{\underset{H}{|}}{\overset{\overset{H}{|}}{C}} - \underset{\underset{OH}{|}}{C} = O$$

ACETAMIDE

$$H - \underset{\underset{H}{|}}{\overset{\overset{H}{|}}{C}} - \underset{\underset{NH_2}{|}}{C} = O$$

71

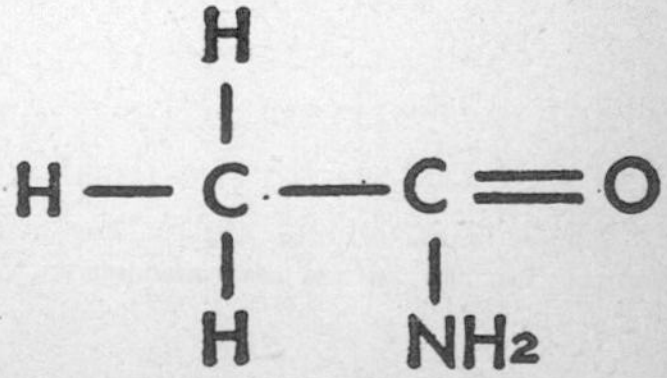

acids derived from unsaturated hydrocarbons. Ordinary vinegar is a carboxylic acid, known as *acetic acid.* Its structural formula looks like the formula for acetaldehyde, except that OH takes the place of the H that is attached to the carbonyl group CO.

Carboxylic acids resemble aldehydes in that both contain the carbonyl group. Carboxylic acids resemble alcohols in that both contain the hydroxyl group. For this reason, in chemical reactions, carboxylic acids behave like a cross between aldehydes and alcohols. They are acids because, when they are dissolved in water, a hydrogen ion may break off from the carboxyl group.

If the OH group in a carboxylic acid is replaced by the group NH_2, the compound becomes an *amide.* The structural formula for *acetamide* is shown next to the formula for acetic acid, to show the relationship between them. In amides, as in carboxylic acids, the carbon that is joined to oxygen is in the third stage of oxidation.

Complete Oxidation

Carbon is in the fourth stage of oxidation in the molecules of the gas *carbon dioxide* (CO_2). In the fourth stage of oxidation, all four pairs of electrons with which carbon forms its bonds to other atoms have been pulled away. Since there are no more pairs to pull away, carbon in the fourth stage of oxidation cannot be oxidized any further.

New Compounds from Old

Carbon compounds can take part in many kinds of chemical reactions in which molecules lose, gain, or reshuffle their atoms to form new compounds. There are several main types of reactions that have been found to take place. One of them is an *addition* reaction, in which a carbon atom which is joined to another carbon atom by more than one bond, breaks one of the bonds so that it can attach itself to one more atom than before. The change from acetylene to ethylene, and from ethylene to ethane are examples of addition reactions. In acetylene each carbon atom is joined to two atoms, in ethylene to three atoms, and in ethane to four atoms.

Another type of reaction is an *oxidation-reduction* reaction. An example is given in the next diagram, which shows how two molecules of acetaldehyde combine with one molecule of water. The oxygen of the water joins one of the acetaldehyde molecules. By changing H to OH, it turns the acetaldehyde into acetic

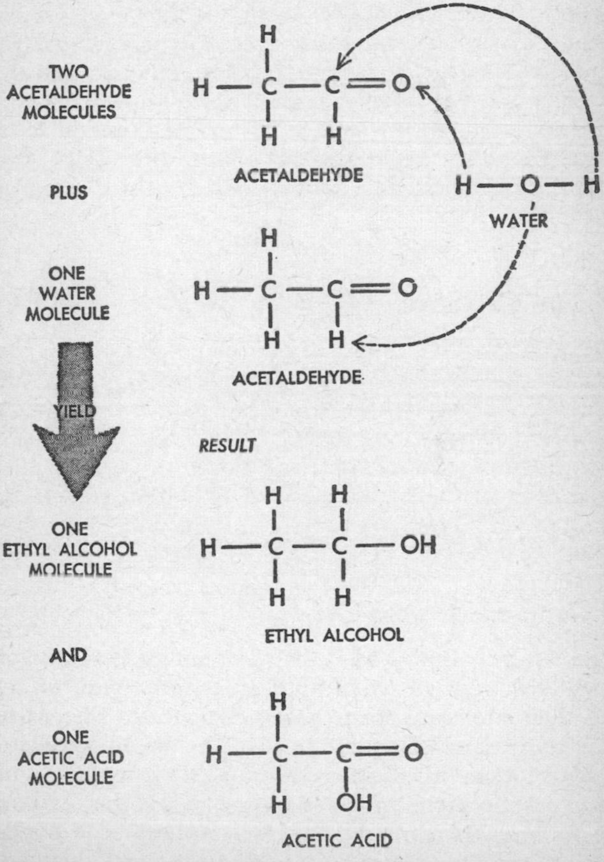

TWO ACETALDEHYDE MOLECULES

ACETALDEHYDE

PLUS

WATER

ONE WATER MOLECULE

ACETALDEHYDE

YIELD

RESULT

ONE ETHYL ALCOHOL MOLECULE

ETHYL ALCOHOL

AND

ONE ACETIC ACID MOLECULE

ACETIC ACID

acid. At the same time, the two hydrogen atoms of the water join the other acetaldehyde molecule. By changing O to OH, and hooking another H on to the C, they turn the acetaldehyde into ethyl alcohol. The change from acetaldehyde to acetic acid is an oxidation, because it carries a carbon atom from the second stage to the third stage of oxidation. The change from acetaldehyde to alcohol is a reduction because it carries a carbon atom from the second stage to the first stage of oxidation. It was also an addition reaction because one of the bonds in a double bond was broken so that a carbon atom joined to only three atoms could be joined to a fourth.

A third type of reaction is a *condensation* reaction, in which, by joining carbon to carbon, two carbon chains are attached to each other to form a longer chain. The diagram below shows how two acetaldehyde

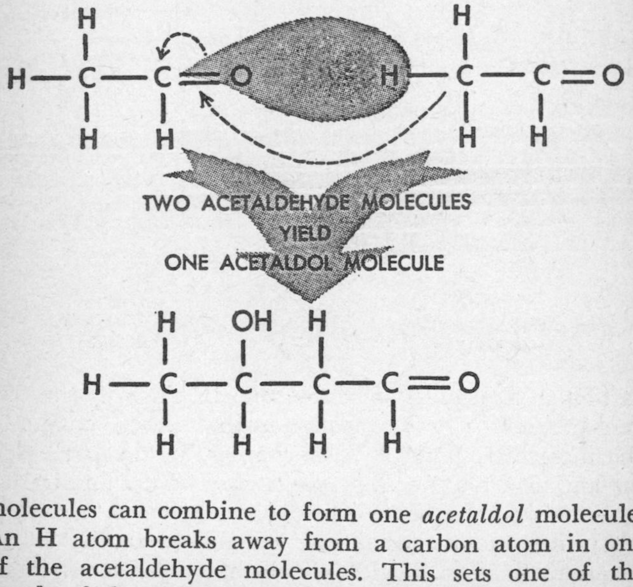

TWO ACETALDEHYDE MOLECULES
YIELD
ONE ACETALDOL MOLECULE

molecules can combine to form one *acetaldol* molecule. An H atom breaks away from a carbon atom in one of the acetaldehyde molecules. This sets one of the bonds of the carbon atom free. The H atom joins the O attached to a carbon atom in the other acetaldehyde molecule. There were two carbon bonds holding the

O, but only one bond is needed to hold the OH, so, here too, one of the carbon bonds is set free. As a result, both molecules have a carbon atom with a free bond, and they use the free bond to join each other to form a single molecule. In this way, two chains that have only two carbon atoms each are joined to form a single chain with four carbon atoms.

A fourth type of reaction is a *polymerization*. Here, too, small molecules combine to form a larger molecule. But while in a condensation one carbon chain is joined directly to another by a carbon-to-carbon bond, in a polymerization the chains are connected by a bridge. The bridge may be an oxygen or a nitrogen atom. For example, three molecules of *formaldehyde* can join to form a single molecule of *paraformaldehyde,* as shown in the diagram. The oxygen in each molecule of formaldehyde uses on of its two bonds to hook on to the

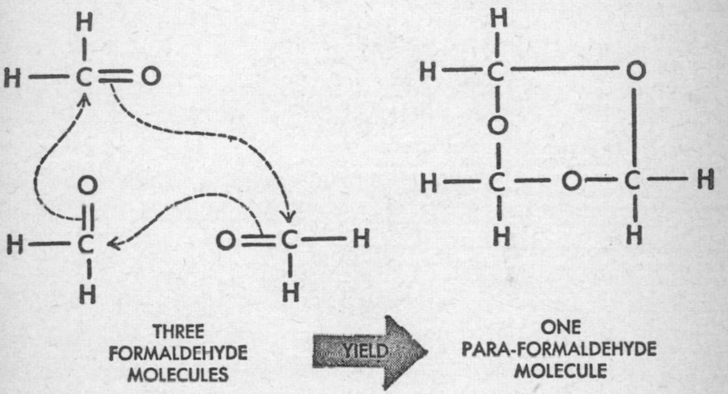

THREE FORMALDEHYDE MOLECULES — YIELD → ONE PARA-FORMALDEHYDE MOLECULE

carbon in a neighboring molecule. In this way the oxygen becomes a bridge between the carbon chains in the molecules. Later in this chapter, in the paragraph on amino acids, we shall see a case of polymerization where nitrogen serves as a bridge.

An important special case of polymerization is the reaction of an acid with an alcohol. The action takes place between the OH group of the acid and the OH group of the alcohol. The two hydrogen atoms and one of the oxygen atoms form a molecule of water. The other oxygen atom serves as a bridge to join the carbon chain of the acid to the carbon chain of the alcohol. When the chains are joined, the molecule of water is squeezed out. What is left is called an *ester*. The diagram on page 77 shows how acetic acid and ethyl alcohol join to form the ester *ethyl acetate* and a molecule of water. Ethyl acetate is found in apples.

Splitting Molecules

In the formation of an ester, two carbon compounds join to form a larger molecule, and, at the point where they join, a molecule of water is squeezed out. This process can be reversed. A molecule of water can split a large molecule into two pieces, meanwhile hooking one of its H's on to one piece, and the remaining OH on to the other piece. This kind of splitting of molecules by water is called *hydrolysis*. The diagram for the formation of ethyl acetate, if read backwards, serves as an example.

The Role of Water

Water plays a very important part in the chemistry of carbon compounds. In the first place, many of them can dissolve in water. When they are dissolved, they are broken up into separate molecules. If different compounds are in the same solution, their molecules are mixed and brought together, so that they can react to form new compounds. In the second place, water itself enters into many of these reactions. We have seen, for example, how a water molecule reacts with two acetaldehyde molecules to oxidize one of them and reduce the other. We have seen how an acid and an alcohol squeeze out a molecule of water when they combine to form an ester. We have also seen how in hydrolysis, water acts to split a large molecule into two smaller ones.

Water got a very early chance to play its part in organic chemistry. When the earth was still young, and the first rains were pouring out of the sky, they car-

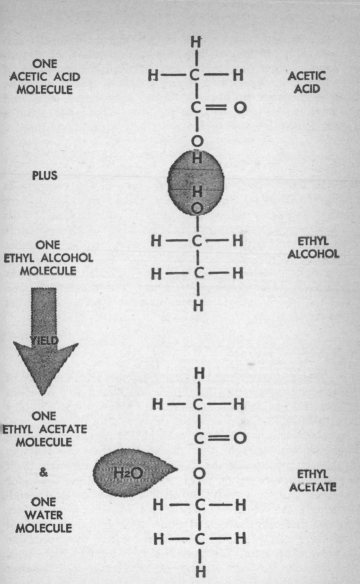

ONE
ACETIC ACID
MOLECULE

PLUS

ONE
ETHYL ALCOHOL
MOLECULE

YIELD

ONE
ETHYL ACETATE
MOLECULE

&

ONE
WATER
MOLECULE

ACETIC
ACID

ETHYL
ALCOHOL

ETHYL
ACETATE

ried into the sea the first organic compounds. These compounds, dissolved in the water of the sea, began to react with each other and with the water itself. Every type of reaction described in this chapter was able to take place in the warm waters of the primitive oceans. With simple hydrocarbons supplying the first carbon chains, these reactions could change them to form alcohols, aldehydes, ketones, and carboxylic acids. Since ammonia was dissolved in the water, too, some of them were changed to amines and amides. By condensation and polymerization, small molecules were joined to make big ones, and by hydrolysis, big molecules were split again into smaller ones. *Out of all this mixing and reacting, complex organic molecules were built up, and these paved the way for the origin of life.*

$$H-\underset{\underset{OH}{|}}{\overset{\overset{H}{|}}{C}}-\underset{\underset{OH}{|}}{\overset{\overset{H}{|}}{C}}-\underset{\underset{OH}{|}}{\overset{\overset{H}{|}}{C}}-\underset{\underset{OH}{|}}{\overset{\overset{H}{|}}{C}}-\underset{\underset{OH}{|}}{\overset{\overset{H}{|}}{C}}-\overset{\overset{H}{|}}{C}=O$$

GLUCOSE

$$H-\underset{\underset{OH}{|}}{\overset{\overset{H}{|}}{C}}-\underset{\underset{OH}{|}}{\overset{\overset{H}{|}}{C}}-\underset{\underset{OH}{|}}{\overset{\overset{H}{|}}{C}}-\underset{\underset{OH}{|}}{\overset{\overset{H}{|}}{C}}-\underset{\underset{O}{\|}}{C}-\underset{\underset{H}{|}}{\overset{\overset{H}{|}}{C}}-OH$$

FRUCTOSE

Sugar, Starch, and Wood

The glycerol molecule, described on page 70 is an alcohol in which an OH group is attached to each of its three carbon atoms. If an H and OH are removed from one of these carbon atoms, and are replaced by a single oxygen atom held by a double bond, the carbonyl group CO is formed. If the carbonyl group is at the end of the carbon chain, the molecule is like an aldehyde. If the carbonyl group is in the middle of the chain, the molecule is like a ketone. But, since the other carbon atoms in the chain still have OH groups

78

attached to them, the molecule remains an alcohol. A molecule like this, which is both an alcohol and an aldehyde, or both an alcohol and a ketone, is called a *simple sugar*. The aldehyde-type sugars are called *aldoses*. The ketone-type sugars are called *ketoses*. Simple sugars may contain as few as three carbon atoms in them, or as many as ten. The most common of the simple sugars found in nature are six-carbon sugars. *Glucose,* the sugar found in grapes, is a six-carbon aldose. *Fructose,* found in the juices of many fruits, is a six-carbon ketose. Structural formulas for a form of each of these sugars are shown on page 78.

Simple sugars may be formed from smaller molecules by polymerization and condensation. For example, if formaldehyde is dissolved in limewater and allowed to stand, the formaldehyde polymerizes to form a mixture of simple sugars. The simple sugars, too, can combine by condensation and polymerization to form larger molecules of three types, *complex sugars, starches,* and *cellulose*. Cane sugar is an example of a complex sugar. Cotton and wood are examples of cellulose. Complex sugars and starches can be broken up easily, by hydrolysis, into the simple sugars of which they are formed. Cellulose can also be hydrolyzed to form sugars, but the process is difficult to carry out. If a simple, inexpensive way of hydrolyzing cellulose is ever developed, chemists will be able to change wood into sugar that can be used for food.

Fats

Because glycerol has three OH groups in it, it can combine with three acid molecules at a time to form an ester. Each OH group combines with the H of the acid's carboxylic group to form a molecule of water. The three molecules of water are squeezed out, and what is left of each acid is joined by an oxygen bridge to one of the carbon atoms of the glycerol. The ester that results is a *fat* or an *oil*. The process is shown in the diagram below. The carboxylic group of each acid is shown, and the rest of the acid, to which the carboxylic group is attached, is represented by a rectangle.

79

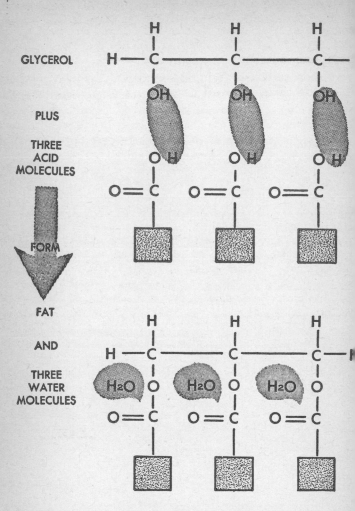

GLYCEROL

PLUS

THREE ACID MOLECULES

FORM

FAT

AND

THREE WATER MOLECULES

The acids which form fats or oils are always acids that have an even number of carbon atoms. If the acids are saturated acids (the fatty acids), the fats formed are *hard fats* which are solid at ordinary temperatures. If they are unsaturated acids, the fats formed are *soft fats* which are liquid oils at ordinary temperatures. The fatty acids found in common fats are usually long chain acids that have from 16 to 20 carbon atoms in them.

Amino Acids

In a carboxylic acid, the carbon at one end of the carbon chain is joined to OH by a single bond and to

GLYCINE

ALANINE

PROLINE

CYSTINE

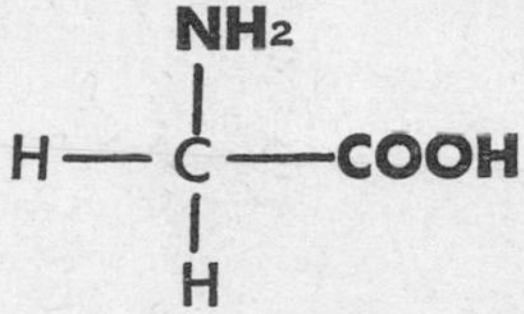

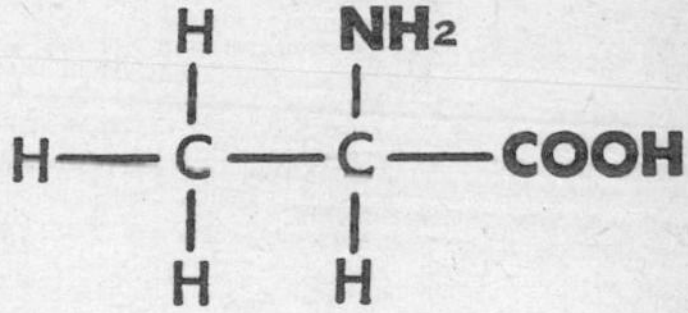

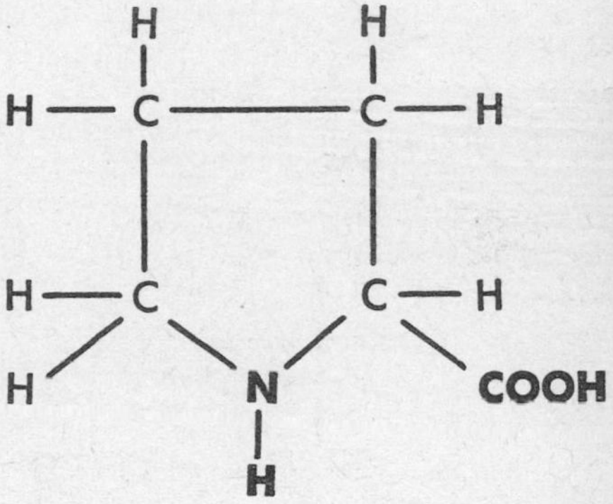

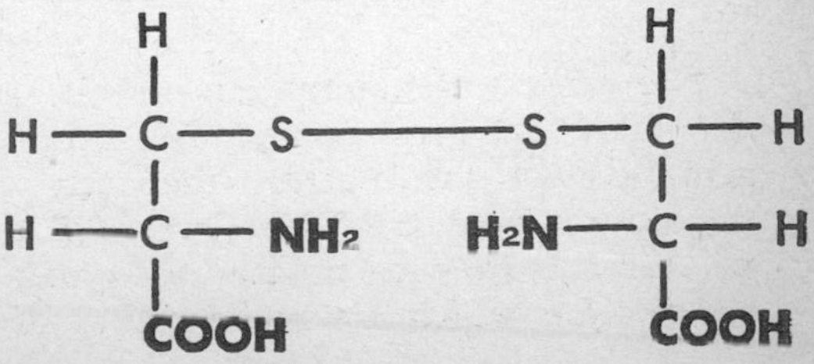

O by a double bond to form the carboxyl group COOH. If a hydrogen atom joined to one of the other carbon atoms in the chain is replaced by the amino group NH_2, the molecule becomes an amine. But since it still has the carboxyl group, it is an acid, too, so it is known as an amino acid. The structural formulas for some amino acids are shown on page 81. *Glycine* and *alanine* have straight carbon chains. *Proline* has a ring structure. The ring is like the pyrrole ring shown on page 65, except that it has been saturated by the addition of hydrogen, so that all the carbon atoms are joined by single bonds. A few amino acids, like proline, have an NH group instead of NH_2. *Cystine* looks like two alanine molecules joined by a bridge of two sulphur atoms that take the place of two hydrogen atoms. Cystine has two carboxyl groups and two amino groups. *Amino acids are important as the building blocks out of which proteins are made.*

In all the amino acids from which proteins are formed, the amino group and the carboxyl group are attached to the same carbon atom. This makes it possible to use a simplified diagram to represent such an amino acid. Instead of showing all the atoms in the molecule, the simplified diagram shows only that part of the carbon chain that has the amino and carboxyl groups. The rest of the molecule, known as the *side chain,* is represented by a rectangle.

Amino acids can combine with each other through their amino and carboxyl groups. The amino group of

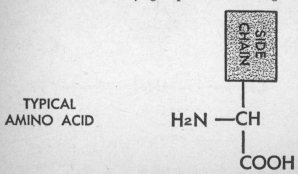

TYPICAL
AMINO ACID

H₂N —CH
|
COOH

82

one amino acid hooks on to the carboxyl group of another, and a molecule of water is squeezed out. The connection between the amino acids is established between the NH and CO that remain. This connection is called a *peptide bond*. The formation of a peptide bond is shown in the diagram below.

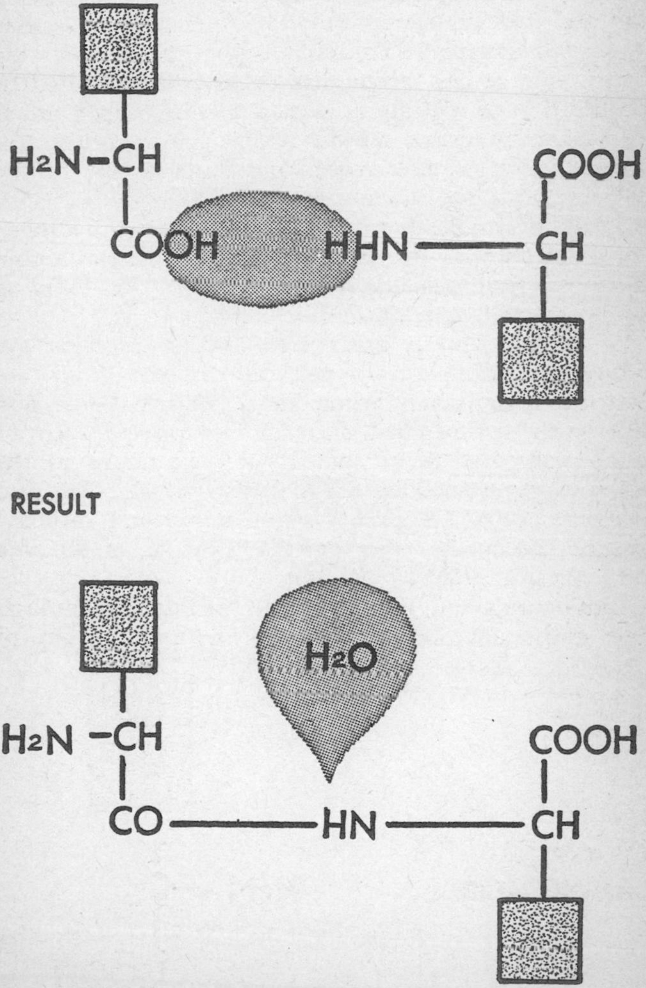

RESULT

We can form a simple picture of the way amino acids join by comparing an amino acid with a person. Think of the amino group as his left hand, and the carboxyl group as his right hand. When two amino acids are joined by a peptide bond, it is like two people holding hands, the left hand of one grasping the right hand of the other. But then each still has a hand free with which he can grasp the hand of another person. This makes it possible to build a chain of amino acids joined by a series of peptide bonds, just as we can form a chain of people holding hands. Such a chain of amino acids is called a *peptide chain*. *All proteins are built out of peptide chains*. By hydrolysis, they can be broken up into the amino acids of which they are formed. There are hundreds of thousands of different kinds of proteins, but they are all built up out of only about two dozen different kinds of amino acids.

Hydroxy Acids

If the amino group in an amino acid is replaced by a hydroxyl group, the result is called a *hydroxy acid*. Just as an amino acid is both an acid and an amine, because it contains both the carboxyl and the amino groups, a hydroxy acid is both an acid and an alcohol, because it contains both the carboxyl and the hydroxyl groups. *Lactic acid*, found in sour milk, and *citric acid*, found in orange juice, are both examples of hydroxy acids. Their structural formulas are shown below.

LACTIC ACID

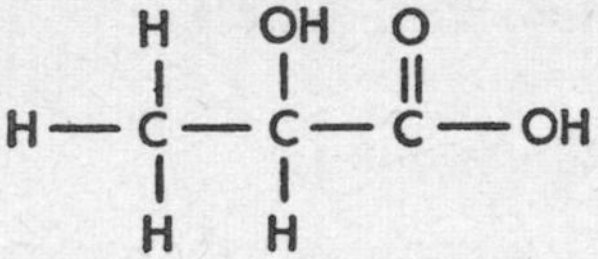

CITRIC ACID

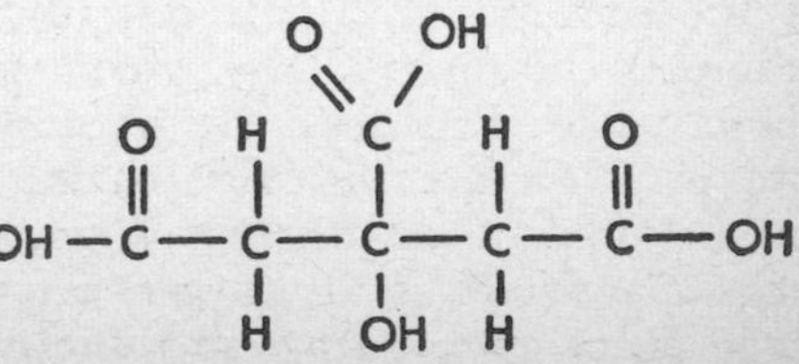

Lactic acid is produced in sour milk by the action of bacteria that cause the sugar in the milk to ferment. It is also produced in muscle cells when they contract. Citric acid plays an important part in the chain of chemical reactions by which living cells burn sugar and fats.

Oxo Acids

Just as amino acids contain the amino group in addition to the carboxyl group, and hydroxy acids contain the hydroxyl group in addition to the carboxyl group, there is a third group of acids that contain the carbonyl group (CO) in addition to the carboxyl group. Acids in this third group are called *oxo acids*. While the carboxyl group makes them behave like acids, the carbonyl group makes them behave like aldehydes or ketones. An example of an oxo acid is *pyruvic acid,* whose structural formula is shown below.

PYRUVIC ACID

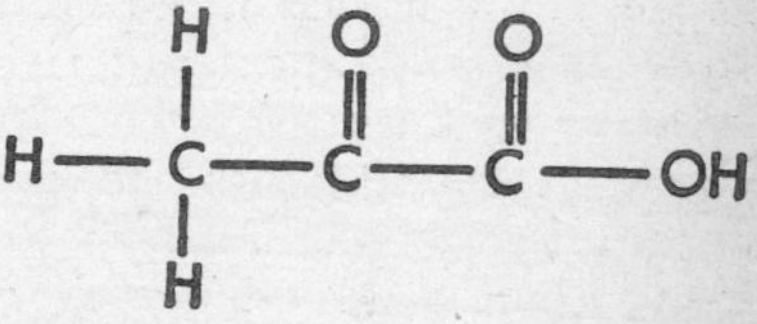

$$H - \overset{\overset{\displaystyle H}{|}}{\underset{\underset{\displaystyle H}{|}}{C}} - \overset{\overset{\displaystyle O}{\|}}{C} - \overset{\overset{\displaystyle O}{\|}}{C} - OH$$

When glucose is fermented it is changed by a series of steps into pyruvic acid, and then the pyruvic acid is converted into ethyl alcohol.

Some Ring Compounds

Some of the compounds that play an important part in life processes are built out of ring structures. For example, chlorophyll, the green pigment that helps plants make their food by photosynthesis, consists of four pyrrole rings, with side chains attached, surrounding an atom of *magnesium. Hemoglobin,* the red pigment that serves as an oxygen carrier in the blood, is made up of pyrrole rings surrounding an atom of iron.

85

Nucleic acids contain four kinds of ring structures called *adenine, thymine, guanine,* and *cytosine.* Thymine and cytosine are like pyrimidine, shown on page 65, but with different side chains attached to the ring. Adenine and guanine are like the double-ring compound purine with different side chains.

On page 78, the simple sugar glucose was represented as an open-chain molecule. The open-chain formula serves to explain its aldehyde character, but it fails to explain other characteristics of the behavior of glucose. Chemists now believe that it is more correct to represent glucose, too, as a ring. The ring is formed by five carbon atoms and an oxygen atom. The sixth carbon atom is part of a side chain attached to the ring. The structural formula for one of the possible glucose rings is shown below.

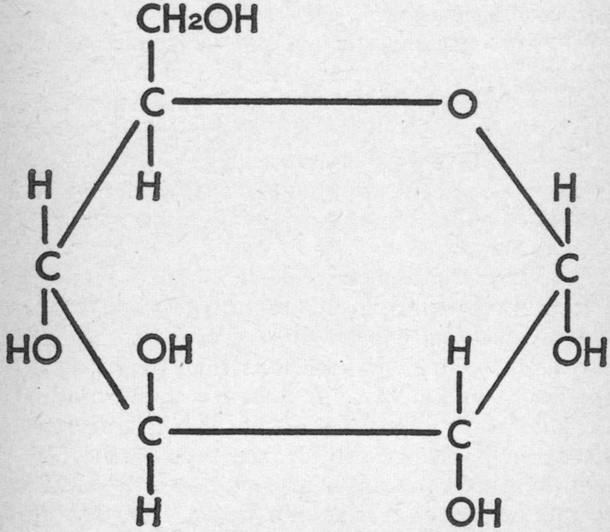

GLUCOSE

CHAPTER V

The Chemistry of Life

Finding Out How

CHAPTER II DESCRIBED some of the chief processes that take place inside living organisms. Using the tools of chemistry, scientists have been studying these processes to try to find out exactly how they work. They have been studying photosynthesis, to find out how plants combine water and carbon dioxide to make sugar. They have investigated the ways in which plants and animals "burn" sugar, starch, and fats to release energy. They have studied the way this energy is used to make muscle cells contract. They have searched for the methods by which amino acids are linked together in peptide chains to form proteins. They have tried to discover the secret of how chromosomes can make copies of themselves.

These investigations into the chemistry of life have led to great progress in several directions. First, more and more facts have been discovered about the chemical structure of the big molecules that are made in living cells. Secondly, there is a better understanding of the chemical processes by which these molecules are built up or broken down. It has been found, for example, that each process is made up of a series of steps that take place one after the other. Thirdly, it has been found that there is a separate enzyme that serves as a catalyst for each of these steps, and many of these enzymes have been identified. Finally, important advances have been made towards understanding how cells capture, store, and use energy.

87

Storage Battery of the Cell

There is a transfer of energy in every chemical reaction. In some reactions energy is released. In others, energy is used up. These two types of chemical reactions are tied together in the cell by the existence of special chemical compounds that serve as the storage batteries of the cell. When energy is captured from sunlight, or released by burning fuel, it is stored in these compounds. When a process takes place in which energy is used up, these "storage batteries" supply the energy that is needed.

The most important of the storage batteries of the cell is a molecule called *adenosine triphosphate,* usually abbreviated as *ATP.* Some other compounds that also serve as storage batteries are built out of ATP. The ATP molecule is made up of three main parts. It contains the double-ring structure adenine that was described on page 86. The adenine is joined to a sugar called *ribose,* and the ribose is attached to three *phosphate groups.* All these parts are coupled to each other in a line, like the cars of a railroad train. ATP can be built up by a step by step process, starting with

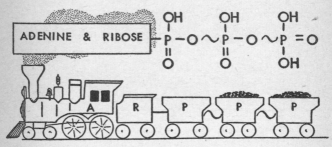

adenine linked to ribose linked to only one phosphate group. To attach a second phosphate group, energy must be supplied to it. The energy is then stored in the bond that joins the second phosphate group to the first one. Chemists call this bond a *high-energy bond,* and they represent it in their diagrams by the wavy line shown in the drawing. The molecule that results is called *adenosine diphosphate,* abbreviated as *ADP.*

To attach a third phosphate group to form ATP, more energy must be supplied. This energy is stored in a high-energy bond between the third phosphate group and the second one.

ATP can transfer some of its stored energy to another organic molecule by giving up to it the third phosophate group together with its high energy bond. The loss of the phosphate group converts ATP into ADP. The molecule to which the phosphate group was transferred can use up the stored energy in carrying out some chemical reaction. When the reaction is over, the phosphate group, stripped of its high energy bond, is released. After that, it can be attached to ADP again to form ATP if energy is supplied once more to be stored in the high energy bond.

Photosynthesis

In green plants, the energy that is stored in the high energy bonds of ATP is supplied by sunlight. The energy of sunlight is captured, and the ATP molecules are built up by processes that are not yet understood.

These processes take place in the *chloroplasts*. The sunlight that enters the cell also serves to break up molecules of water, separating hydrogen atoms from the oxygen with which they are combined. The oxygen is released into the air, and the hydrogen is combined with carbon dioxide to form glucose.

In the year 1955, scientists finally worked out the details of the process by which glucose is built. The process takes place in a series of nine steps. ATP and another similar compound provide the energy for carrying out the process, and then this energy is stored in the glucose. At each step of the process a special enzyme serves as catalyst.

The series of steps in photosynthesis is shown in the diagram below. The Roman numerals show the order of the steps. The starting point of the process is a molecule of a five-carbon sugar (one with five carbon atoms in the molecule). A supply of these molecules was in the seed from which the green plant originally grew. In the first step of the process, a carbon atom from carbon dioxide is joined to the five-carbon sugar to form a six-carbon compound. In the second step, the six-carbon compound is split to form two molecules of a three-carbon acid. In the third step, hydrogen from water is joined to the two molecules of acid to

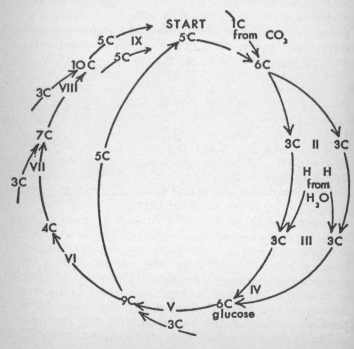

change them into two molecules of a three-carbon sugar. In the fourth step, the two molecules of three-carbon sugar are joined to form a molecule of the six-carbon sugar glucose. In the fifth step, the six-carbon sugar is joined to a three-carbon sugar to form a nine-carbon compound. In the sixth step, the nine-carbon compound is split into a five-carbon sugar and a four-carbon sugar. The five-carbon sugar can be used for starting the process all over again. The four-carbon sugar goes on through a seventh step in which it is combined with a three-carbon sugar to form a seven-carbon sugar. In the eighth step, the seven-carbon sugar is joined to a three-carbon sugar to form a ten-carbon compound. In the ninth step, the ten-carbon compound is split into two molecules of five-carbon sugar which can be used to start the process all over again.

Because the process is a chain of steps that ends where it begins, it is called a cycle. The cycle is repeated over and over again. The plant started life with a large number of molecules of the five-carbon sugar, and each serves as the starting point of a cycle, so that there are many cycles of the same kind taking place side by side in each cell. Notice that in each cycle, molecules of three-carbon sugar are needed in steps five, seven, and eight. These molecules are provided by step three of neighboring cycles.

It may look as though the whole process is useless as a way of making glucose, because the glucose produced by step four is used up in the next steps in order to produce more molecules of the five-carbon sugar. But this is not so. What happens may be thought of as proceeding in this way: Suppose the cell begins with five molecules of the five-carbon sugar, and only a single cycle goes on at a time. Suppose that the first three cycles stop at step three. This uses up three of the molecules of the five-carbon sugar to produce six molecules of the three-carbon sugar. Then the fourth and fifth cycle can proceed through all nine steps, using these six molecules and the remaining two molecules of the five-carbon sugar, and producing six molecules of the five-carbon sugar. Then, if the sixth cycle goes on only through the fourth step, one of these molecules of five-carbon sugar is turned into glucose. The final outcome of the six cycles is that the five molecules of five-carbon sugar with which the cell started are all restored, and *in addition* a molecule of glucose has been built.

What really happens is not merely a repetition of a single cycle, sometimes stopping after three steps, sometimes after nine steps, and at other times after four steps. Actually there are many cycles going on at the same time, and they are not all in step. When one cycle reaches steps five, seven, or eight, where molecules of the three-carbon sugar are needed, it gets them from other cycles that have just completed the third step. All cycles that go on beyond the third step make glucose. In some cases the glucose is used to finish the cycle. In other cases the glucose is withdrawn to take part in other processes that are going on in the cell. It may be burned to supply energy, or it may be built into a molecule of a complex sugar, starch or cellulose. The various processes that go on side by side in the cell are all co-ordinated, so that the materials needed in one process are supplied by the products of other processes, and they are supplied *in the right amounts and at the right time*.

Burning Fuel

In the process by which a glucose molecule is built out of water and carbon dioxide, energy is transferred from ATP to the glucose molecule. This energy remains hidden in the glucose and in the complex sugars and starches that are made out of glucose. The energy is released again when these fuels of living cells are "burned." But it is not released in the form of heat, the way it is when fuel is burned in an ordinary fire. It is released in the form of chemical energy that is stored again in the high energy bonds of ATP molecules. Then it can be withdrawn from the ATP molecules whenever the cell needs it again for some other purpose, like contracting a muscle or building a protein molecule.

When a cell uses a complex sugar molecule or a starch molecule as fuel, first it uses hydrolysis to break the molecule up into molecules of simple sugar. Then, in some cases, the simple sugar is burned by anaerobic process like fermentation, which does not require oxygen. Yeast cells, for example, get their energy by fermenting the simple sugar, glucose, converting it into ethyl alcohol. The process takes place in fourteen steps, each carried out with the help of a special enzyme. The effect of the first twelve steps is to break up each six-carbon molecule of glucose into three-carbon molecules of pyruvic acid. In the thirteenth step, the pyruvic acid is converted into acetaldehyde and carbon dioxide. In the last step, the acetaldehyde is changed to ethyl alcohol. In the first and third steps, energy has to be supplied to the process. The energy is delivered in each case by one ATP molecule that gives up one of its phosphate groups and becomes ADP as a result. But in the eighth and eleventh steps the process releases energy. At each of these steps enough energy is released to convert two ADP molecules into ATP molecules. As a result of the entire process, for every glucose molecule that it ferments, the cell gets two more ATP molecules than it started with.

Organisms that burn their fuel by respiration, in which oxygen is used, get a better energy return for their efforts. The aerobic oxidation of glucose begins by splitting it into pyruvic acid molecules, as in fermentation. But after that it proceeds in a different way. A carbon atom is chopped off from each three-carbon molecule of pyruvic acid and is combined with oxygen to form carbon dioxide. A two-carbon molecule of acetic acid is left. Each acetic acid molecule then enters into a cycle of changes known as the *citric acid cycle*. The two-carbon acetic acid molecule combines with a four-carbon molecule known as *oxaloacetic acid* to form the six-carbon molecule of citric acid. The citric acid molecule, as it proceeds from step to step in the cycle, is changed into other compounds. In the course of these changes, it combines with four atoms of oxygen, and loses two atoms of carbon. The final result of all these changes is a molecule of oxaloacetic acid, ready to combine with another molecule of acetic acid and start the cycle all over again. The conversion of pyruvic acid into acetic acid releases enough energy to form three phosphate bonds that change ADP to ATP. The citric acid cycle releases enough energy to form twelve more. Since each molecule of glucose can form two molecules of pyruvic acid, the aerobic oxidation of the whole molecule releases enough energy to convert over thirty molecules of ADP into ATP.

Each step in the aerobic oxidation of glucose is controlled by an enzyme. The enzymes that do the job are in the rod-shaped mitochondria that float in the cytoplasm of the cell.

Organisms also burn fat as a fuel. First the fat is hydrolyzed to split each fat molecule into a molecule of glycerol and three fatty acid molecules. Then, with the help of *coenzyme A,* a compound that contains ATP, each fatty acid is chopped into two-carbon fragments to form molecules of acetic acid. Then the acetic acid molecules are burned in the citric acid cyle. While ATP has to be supplied to the process of splitting the fatty acids, the citric acid cycle more than makes up for it by the ATP molecules that it produces. In the oxidation of fats, for every ATP molecule that is used up, about one hundred new ones are formed.

Moving Muscles

The energy that is stored in ATP can be used to make a muscle move and do work. Muscle cells contain fibers of a protein called *actomyosin.* When ATP molecules pass their stored energy on to the actomyosin fibers, the fibers contract and the muscle moves. The ATP molecules that set off the contraction are produced when they are needed by drawing on the energy that is stored in the muscle cells in the form of a starch called *glycogen.* By a process of anaerobic oxidation, the glycogen is converted into lactic acid. The energy that is released changes ADP to ATP. Then the ATP triggers the contraction of the actomyosin fibers. A muscle becomes tired when its glycogen supply is used up. Its strength is restored by a period of rest. During the rest period, fuel is burned to release energy. The energy is used to change lactic acid back again into glycogen. Then the muscle is ready to contract again. Each step in these processes is controlled by specific enzymes.

Building Carbon Chains

In the oxidation of fats we saw coenzyme A, a relative of ATP, working as a molecule chopper. It served to break the carbon chains of fatty acids into fragments containing only two carbon atoms each. But coenzyme A can also serve as a molecule builder. Starting with two-carbon fragments, it can join them together to form a long carbon chain. A coenzyme A molecule attaches itself to each two-carbon fragment to form a larger molecule called *acetyl coenzyme A*. Then two of these larger molecules join and squeeze out one of the coenzyme A molecules. In the process, the two-carbon fragments are linked to form a four-carbon chain. By a similar process, the four-carbon chain can be linked to a two-carbon fragment to form a six-carbon chain. Through a series of such steps, carbon chains can be lengthened, branches can be hooked on to them, or they can be looped to form rings.

Proteins

The chemical compounds that are most characteristic of living things are proteins. In the human body, about half of the dry matter is protein. The family of protein compounds has a multitude of members. There may be as many as one hundred thousand different kinds of proteins in a single body. Each has a special job to do, and its chemical nature fits it for its job. The protein actomyosin, found in muscle fibers, has the job of contracting, to enable the muscle to do work. *Collagen* in the skeleton helps to give it its strength. The enzymes which control the speeds of chemical reactions in the body are all proteins. *Hormones,* the chemical regulators manufactured by glands in the body, are also proteins. When the body is invaded by germs and attacked by their poisons, proteins known as *antibodies* come to the body's defense.

All proteins are made up of amino acids joined together in a peptide chain, as described on page 83. The next diagram, of a short peptide chain, shows how the amino acid side chains are linked by CO and NH groups. There are only about two dozen different kinds of amino acids found in nature. Arranging the amino acids in a peptide chain is like arranging letters of the alphabet to spell a word. By choosing different groups of letters and arranging them in different ways we can

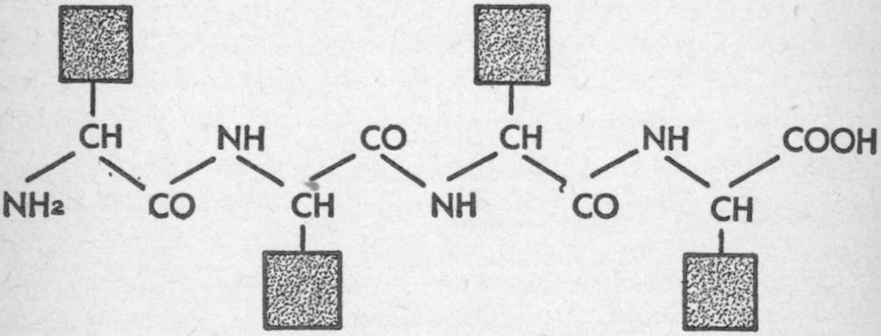

NH_2 CH NH CO CH NH COOH
CO CH CO NH CO CH

spell out different words. The same letter can be used more than once in one word, so the number of letters in a word can be more than the number of letters in the alphabet. In the same way, different groups of amino acids arranged in different ways form different proteins. That is why so many different kinds of proteins are possible. When amino acids are joined, energy is needed to form the peptide bond. Here again the energy is supplied by ATP.

Protein molecules are large and heavy. The smallest known protein molecule is as heavy as thirteen thousand hydrogen atoms. The largest known protein molecules have about fifty thousand amino acids in them and weigh as much as about ten million hydrogen atoms.

Its peptide chain structure makes a protein molecule like a thread. But the thread isn't always stretched out. In some proteins the thread is coiled up or tangled. We saw on page 53 that a hydrogen atom that is in a molecule still has an extra little hook, the hydrogen bond, by which it can attach itself to a neighboring oxygen, nitrogen or fluorine atom. In many protein molecules, hydrogen atoms in one part of the peptide chain hook on to oxygen or nitrogen atoms in another part of the chain. In this way the loops and coils of a tangled molecule are held in place.

To find out the chemical structure of a protein, it is necessary first to identify the amino acids that are in it, and then to find out the order in which they are arranged. This was done successfully for the first time in 1954, when the structure of the molecule of *insulin* was finally worked out. Insulin is a hormone manufactured by the pancreas. It plays a part in the oxidation of fatty acids. There are seventeen different kinds of amino acids in the insulin molecule. They are arranged in two peptide chains that lie side by side. There are 21 amino acids in one chain, and 30 in the other. Cystine is one of the amino acids in the molecule. The diagram on page 81 shows that cystine is made up of two identical parts joined by a sulphur bridge. Each part contains a carboxyl group and an amino group with which it can form peptide bonds. This makes it possible for one part of the cystine to lie in one of the peptide chains, while the other part is in the other peptide chain. The sulphur bridge between the two parts holds the two peptide chains together. There are two of these sulphur bridges joining the two chains in the insulin molecule.

Anagrams and Patterns

In the game of anagrams, letters are spread out face down on a table. Each player in rotation turns one of the letters face up, and then using the letters turned up, he tries to form a word. What words he can make depends on what letters are turned up by the players.

The construction of proteins in a living cell has some resemblance to a game of anagrams. It begins with a supply of amino acids, just as the game of anagrams begins with a supply of letters. If the organism is autotrophic, it makes its own amino acids. If it is heterotrophic, it gets most of its amino acids from the food it eats. Then, in the process of making proteins, groups of amino acids are picked out and joined together. But here the resemblance to anagrams ends. While in anagrams the choice of letters is accidental, in the construction of proteins the choice of amino acids is not. The cells in the pancreas, for example, have a way of picking out exactly those amino acids that are needed to make insulin, and always manage to arrange them in the right order. This suggests that the cells have built-in patterns that they can follow to make particular proteins. It is thought that these patterns are provided by the nucleic acids in the cell.

Molecules that Copy Themselves

There are two types of nucleic acids in cells. One type, called *desoxyribonucleic acid,* and usually abbreviated as *DNA,* is found only in chromosomes. The other type, called *ribonucleic acid,* and abbreviated as *RNA,* is found mostly in the cytoplasm, although there is some in the nucleus too. Both DNA and RNA play a part in the cell's work of putting amino acids together to make proteins. But the most interesting job that they do is to *make exact copies of themselves.* The power to do this is the basis of a cell's ability to reproduce by dividing into two complete cells.

A DNA molecule is built out of long chains, just as proteins are made up of chains. In proteins, the units that are joined together to form a chain are amino acids. In DNA, the units are structures that look something like ATP. Each DNA unit consists of a ring structure attached to a five-carbon sugar, which is attached to a phosphate group. When the units are joined in a

DNA chain, the sugar molecules and phosphate groups form alternate links, and the ring structures stick out as side chains, one on each sugar molecule. There are only four ring structures that appear in DNA. Thymine and cytosine contain a single ring of the pyrimidine type. Adenine and guanine contain a double-ring structure of the purine type. There may be as many as three thousand units in one DNA chain. RNA molecules have the same kind of structure as DNA, but they contain a different kind of sugar.

A DNA molecule consists of two chains spiraling around each other like the twisted strands of a rope. Each side chain of one strand is joined to a side chain of the other strand by a hydrogen bond, and these linked pairs hold the two strands together. In these linked pairs, adenine is always joined to thymine, and guanine is always joined to cytosine. Apparently the side chains that are partners in a pair have a special attraction for each other which causes each one to join only the one type of side chain for which it has a preference.

The special attraction that each side chain has for its preferred partner suggests a way in which DNA molecules may be able to copy themselves. The two strands which are twisted around each other in the molecule unwind. The hydrogen bonds are broken, and the strands separate. In the fluid that surrounds the strands there are four kinds of DNA units. Each side chain on the separated strands attracts to itself that particular unit that contains the side chain it prefers. It hooks on to the side chain of this unit by means of a hydrogen bond, and holds it in place. When all the places alongside a strand are filled by the captured units they are joined to each other to form the phosphate-sugar chain. Then there are two double strands exactly like the one with which the process began. This process is an example of how a complex molecule can serve as a pattern for building another molecule. It is now thought that a similar process may take place in the construction of proteins.

Natural But Complicated

There are many things about the chemistry of life that are not yet understood. But as scientists continue their investigations, our knowledge about it steadily grows. This chapter summarizes some of the recent discoveries that have added to our knowledge. These discoveries tend to confirm the view that life processes can be explained as the natural outcome of the operation of the laws of physics and chemistry. But they also show how extremely complicated these processes are. The life process of even the simplest one-celled organism is made up of a large number of chemical reactions many of which take place side by side while others follow each other in time. All are regulated so that one process feeds another. Special enzymes are there to serve as regulators, and there is a special machinery for capturing, storing, and transferring the energy that is needed to keep the process going. An explanation of the origin of life will have to answer the question, "How did this complex, organized process develop spontaneously out of the simple chemicals and the unorganized chemical processes that preceded it in the primitive seas billions of years ago?"

CHAPTER VI

Floating Specks

Molecules Are Sticky

THE ATOMS IN a molecule are held together by electrical forces. The nucleus of each atom has a positive charge, and the electrons that surround the nuclei have negative charges. These opposite charges attract each other. The force of attraction serves as a cement binding the parts of the molecule to each other. But these parts are not rigidly fixed. They are in motion, and there are spaces between them. The electrical forces leak through these spaces and bring their cementing action to the surface of the molecule. As a result, the molecule is sticky, and it will cling to other molecules that are close to it. When large groups of molecules are held together by their stickiness they form liquid droplets or solid particles.

When solid particles are mixed with water, they are surrounded by the dance of the water molecules. The pushing, jostling crowd of dancing molecules tends to draw molecules out of the particles and sweep them along in the dance. This creates a tendency for the molecules in the particles to separate and scatter. At the same time, the stickiness of the molecules tends to keep them together. So, while some molecules break away from the particles to join the dance, others leave the dance to rejoin the particles. Finally, a balance between the two tendencies is reached, like the balance between diners and dancers at a crowded night club. People leave the tables to join the dance, while others stop dancing and rejoin the clusters of people seated at the tables. While it is not the same people dancing all the time, the number of people dancing is about the same, and depends on how many the dance floor can hold, *and how strong their interest is in dancing*. There are some chemical compounds, like table salt, whose molecules have a strong tendency to join the dance. When particles of these compounds are mixed with water, the molecules scatter completely or *dissolve*. We say that these compounds are *soluble* in water. There are other compounds, like sand, for example, whose molecules have only a weak tendency to join the dance. Nearly all of them are wallflowers who cling together in the solid particles, and so the solid particles do not dissolve. We say that these compounds are *insoluble*.

Too Small to Sink

If a teaspoon of table salt is stirred in a glass of water, the grains of salt dissolve. The scattered molecules of salt dance around with the molecules of water, and are invisible. If a teaspoon of sand is stirred in a glass of water, the sand does not dissolve. The grains of sand do not join the dance of the molecules. Each grain is very large compared to the moving molecules, and many of them collide with it. But while a large number of collisions push it from one side, an equal number of collisions push it from the other side. So the pushes cancel out and do not make the grain of sand move. But sand is heavier than water. Its weight does make it move, and it sinks through the water. The grains of sand settle out on the bottom of the glass.

The outcome is different if the sand is ground into a very fine powder before it is mixed with the water. A particle of the powder is so small that only a few molecules collide with it at a time. While a molecule pushes it from one side, there may be no opposing push from the other side. The pushes do not cancel out, and they do make the particle move. The weight of the particle tries to pull it down, but it gets pushed up often enough to keep it from settling. It remains floating in the water, tossed this way and that by the dancing molecules, like the ball in a game of soccer. When particles are small enough to be kept afloat in this way, they are called *colloidal particles*, and the mixture of floating specks and water is called a *colloidal solution*.

To measure small particles, a tiny unit of length

known as an *angstrom* is used. There are about 250 million angstroms in an inch. Colloidal particles are between ten angstroms and five thousand angstroms in width. Because they are so small, they are invisible under an ordinary microscope, and can pass through the pores of an ordinary paper filter. A colloidal particle may contain as few as one thousand atoms and as many as one billion.

Colloidal particles may be clusters of small molecules, or they may be single molecules that contain many atoms. *Protein molecules, made up of long chains of amino acids, are large enough to be colloidal particles.* In protoplasm, the material of living cells, proteins and other tiny particles are floating in water. *So protoplasm is a colloidal solution.* By learning how colloidal solutions behave, scientists discovered the explanations for some of the things that happen in protoplasm. *They also found clues to how colloidal solutions that could have formed in the sea two billion years ago might have paved the way for the origin of life.*

Small Particles, Big Surface

If a particle is broken into pieces, new surfaces are formed along the breaks. These new surfaces are in addition to the surface of the original particle. So the small pieces have more total surface than the single piece from which they were formed. The smaller the pieces are made, the greater the surface becomes. Because of this fact, a small amount of a substance broken up into a large number of colloidal particles has a tremendous amount of total surface. For example, if a lump of sugar is one centimeter wide (slightly less than half an inch), its surface area is only about one square inch. But if it is broken up into a thousand billion billion pieces of the same cubical shape, the pieces are colloidal particles, and their total surface area is over seven thousand square yards, or almost an acre.

Sticky Powders

Because very small particles have such a large total surface, a great number of their molecules are on the surface. The combined stickiness of the surface molecules makes each particle very sticky. For this reason very fine powders tend to cling to anything they touch. Ladies make use of this fact when they powder their noses. We all make use of it when we write on paper by rubbing carbon powder over it from the point of a pencil.

The stickiness of small particles surrounded by a gas or liquid makes them pull passing molecules onto their surfaces. When this happens we say that they have *adsorbed* the molecules. When particles adsorb molecules they show a preference for some molecules over others. Powdered charcoal, for example, attracts and holds the poisonous mustard gas better than it holds oxygen. That is why it is used in gas masks. When air that contains poison gas flows through the gas mask, the charcoal removes the poison gas from the air, but lets the oxygen pass on into the lungs of the man wearing the mask.

Their power to adsorb special molecules explains why many colloidal particles are effective as catalysts for the chemical reactions between these molecules. When the particles pull the molecules on to their surfaces, they bring them into close contact with each other, and this increases the speed of the chemical reaction.

Growing Slowly by Leaps and Bounds

Colloidal particles in a colloidal solution are pushed this way and that by the molecules dancing around them. As a result they leap and bound about in the solution. From time to time they collide. But then their stickiness tends to make them cling to each other, and the small particles gradually grow into larger particles. The solution becomes cloudy, and finally the particles may grow large and heavy enough to settle out. This process is called *coagulation*.

While collisions and the stickiness of the colloidal particles tend to make them coagulate, there are other forces that tend to prevent coagulation. Some colloidal particles are ions, and therefore have an electrical charge. Others adsorb ions from the solution, and in this way, they too become charged. When the colloidal particles all carry the same charge, they repel each other. The repulsion prevents them from sticking to each other to form larger particles. The solution remains stable and clear, and the particles do not settle out. However, if an electric current is passed through the solution, or ions with the opposite charge are added, they can neutralize the charge on the particles and make them coagulate.

A Skin of Water

While the stickiness of colloidal particles makes them cling to each other, it may also make them cling to the molecules of water that surround them. There are some that have no preference for water, and will not adsorb the water molecules. They are called *hydrophobic,* or water-fearing, particles. But there are others that do have a preference for water. They are called *hydrophilic,* or water-loving. The hydroxyl group, the amino group, and the carboxyl group in organic compounds have an attraction for water. So colloidal particles that contain these groups are hydrophilic. When hydrophilic particles are in a solution, they adsorb molecules of water. The water molecules clinging to the surface of each particle form a skin around it. Then when the particles collide, the skin keeps them from sticking to each other. In this way the liking that hydrophilic particles have for water serves as another way of preventing coagulation. But if, by some chemical change, their attachment to water is broken, the water skins break up and the particles begin to cling to each other and grow.

Balls and Threads

Colloidal particles may have many different shapes. Some are ball-shaped, while others are long and thin, like tiny threads. Ball-shaped particles are less likely to cling to each other than thread-shaped particles. So a solution of ball-shaped particles flows freely, like water, while a solution of threadlike particles is more *viscous* or slow-moving, like glue.

Protein molecules are long peptide chains. Some are stretched out fairly straight, and are threadlike. Others are wound up in a ball, with the coils linked together by hydrogen bonds. Examples of threadlike proteins are collagen, found in skin, sinews, and bone; myosin, found in muscle; and *keratin,* found in nails, horn, and hair. The threadlike shape of these proteins allows them to build up structures which give the tissues they are in definite shape. Blood proteins are ball-shaped. Their roundness permits the blood to flow freely.

Water in a Cage

When colloidal particles in a solution coagulate, they do not always settle out. If the particles are threadlike and are crowded together, they crisscross against each other to form a cagelike structure resembling a jungle gym in a playground, but with irregularly shaped compartments. Water is trapped in each little compart-

ment and is prevented from flowing. As a result the whole solution becomes stiff and jelly-like. A colloidal solution in this jellied state is called a *gel.* The flowing state from which it is formed by coagulation is called a *sol.*

Sols are turned into gels in different ways, depending

106

on what kind of particles they contain. For example, fruit jellies are a colloidal solution of the protein called *pectin*. They are made to gel by the addition of sugar. Egg white contains the protein *albumin* in which each peptide chain is wound up in a ball. Boiling an egg breaks the hydrogen bonds that bind the coils of the chain to each other. The chain unwinds as a thread. Then the threads coagulate to form the stiff white jelly of a boiled egg. Blood contains the coiled protein *fibrinogen*. Blood clots to form a jelly when fibrinogen unwinds and the threads coagulate. A solution of *gelatin* forms a jelly when it is cooled. Cooling slows down the movements of the gelatin threads, and when they are crowded enough and slow enough, they catch on to each other to form a network of cages.

A Sieve of Jelly

A gel that is thin and sheetlike is called a *membrane*. In a membrane, the network of crisscrossing threads resembles a sieve. The spaces surrounded by the threads are like the holes of the sieve. If the membrane is surrounded by liquid, small molecules in the liquid will pass freely through these spaces, but large molecules will be stopped by the threads. If there are large molecules on one side of the membrane, and only small ones on the other, the small molecules will flow through to where the large ones are. Both large and small molecules will try to flow back, but the large ones will not get through. As a result, fewer molecules will get back than came through in the first place, and the liquid gradually oozes over to where the big molecules are. This one-way flow of a liquid through a membrane is called *osmosis*.

Sol and Gel in the Cell

The protoplasm in a living cell is a colloidal solution containing proteins, fats, sugars, starches, salts, and other chemicals. The electrical and chemical conditions are different in different parts of the cell, and they keep changing all the time. Particles pick up and lose electrical charges. Atoms join in chemical partnerships and then separate again. Colloidal particles adsorb molecules and then release them. Threadlike particles coagulate and then separate again. Because of all this activity, one part of the protoplasm may stiffen from sol to gel, while another part loosens from gel to sol. Permanent structures like the nucleus and the mitochondria are surrounded by membranes, but there are also temporary membranes that form and break up. When the colloidal particles adsorb or release molecules, and when the membranes cause a one-way movement of molecules, they help to regulate the flow of materials within the cell, and control the timing of the complex chemical changes of the life process.

Colloidal Clusters

Under special conditions, in a colloidal solution that contains two or more different kinds of particles, the particles may come together without coagulating, and will form clusters called *coacervates* instead. In a coacervate, particles with opposite charges are drawn close to each other by the electrical force between them, but the water skins that surround the particles prevent them from coming into direct contact. The cluster as a whole is surrounded by a larger skin of water that separates it from the fluid in which it floats. Molecules pass through this skin, so that there is a flow of chemicals into and out of the cluster. Some are adsorbed by the particles, while others are released. As a result of this activity, the forces that hold the cluster together may be either strengthened or weakened. If they are weakened, the cluster breaks up and the particles scatter. Excess water inside a cluster sometimes collects in a little bubble, like vacuoles inside a living cell.

Coacervates form easily in colloidal solutions that contain a variety of large molecules. *The warm waters that covered the earth over two billion years ago were colloidal solutions of this kind. We shall see in Chapter VII how coacervates that could have formed in these waters may have served as an intermediate stage in the development from simple organic compounds to the complex chemical organization known as protoplasm.*

108

From Death to Life

The Simple Is Complex

THE STORY OF biological evolution describes how, through gradual changes and natural selection, more complicated organisms developed from simpler ones. This story begins with the simplest and oldest organisms, like bacteria and blue-green algae. But even these simplest organisms are complex chemical systems in which complicated organic compounds take part in delicately balanced chains of chemical reactions. The problem for students of the origin of life is to explain how these complex chemical systems developed from the simple compounds and processes that existed before them. In the story of the origin of life, the organisms that are the beginning of biological evolution are themselves the end of a long process of chemical evolution. This chemical evolution began with the origin of the earth as a planet.

It is possible to estimate the amount of time during which this chemical evolution took place. We know that blue-green algae already existed two and a half billion years ago, because fossil algae have been found in rocks that are two and a half billion years old. We know, too, from evidence in the rocks, and from clues in the sun and stars, that the earth is about four and a half or five billion years old. This allows a period of about two billion years during which the chemical evolution took place that led to the origin of life.

How the Earth Was Born

There are two different theories of how the earth was formed. According to one theory, a star that passed close to the sun pulled a large cigar-shaped mass of hot gas out of the sun. As the mass of gas cooled and contracted, it split into several pieces and each piece became one of the planets, the earth among them. As the young earth cooled some more, part of the hot gas condensed into liquid and collected at its center. According to the other theory, the planets and the sun were formed at the same time from whirling clouds of dust and gas in space. Attracted to each other by the force of gravitation, small particles collided to form larger particles. The larger particles attracted the smaller ones around them, and grew in size. Finally most of the dust and gas in our part of space were concentrated in several large spheres. In each sphere the particles, pulled by their own weight, pressed in toward the center and made the sphere contract. The colliding, pushing and squeezing generated heat, and the spheres began to grow warm. The largest of the spheres became so hot, it began to glow. This sphere became the sun. The sphere that became the earth was never hot enough to glow, but it did become hot enough to melt all its solid particles, so that it became a liquid ball surrounded by gas. Although the two theories give different pictures of how the earth began, they both lead to the same conclusion: there was a time far back in the earth's early years when it was much hotter than it is today, and when it was a liquid ball surrounded by hot gas.

The Chemicals in the Earth

The two theories also lead to the same conclusion about the chemical nature of the young earth. According to both theories, the earth was formed from the same stuff as the sun. So, by finding out what chemicals there are in the sun, we also find out what chemicals there were in the earth when it was formed. Astronomers have identified the chemicals that are in the sun from clues that they find in sunlight. Additional clues have been obtained from the study of the stars, and from chemical analysis of meteorites, solid particles that continue to fall in on the earth from surrounding space.

From laboratory experiments, chemists know how the chemical elements and the compounds they form behave under different conditions of temperature and pressure. By combining this information with estimates

110

of what conditions were like on the young earth, they have figured out the physical and chemical changes that probably took place as the earth began to cool.

The Earth Forms a Crust

The liquid ball in the young earth included most of the earth's iron. Since the iron was heavier than the other liquids with which it was mixed it sank toward the center. The other part of the ball was made up mostly of oxides and silicon, aluminum, magnesium, calcium, sodium, and potassium. These oxides were mixed and combined with varying amounts of iron. As the earth cooled, the outer part hardened to form the solid mantle and crust. The iron core at the center continued as a liquid to the present day.

The Changing Atmosphere

The earth's first atmosphere was made up mostly of hydrogen and helium, but also included methane, ammonia, hydrogen sulphide, and water vapor. Because of the high temperature, the molecules of gas were moving about rapidly. The molecules that weighed the least, like hydrogen and helium, were moving fastest. Many moved fast enough to break away from the pull of the earth's force of gravity, and they escaped into space. Gradually the entire original atmosphere was lost, the lightest molecules wandering away first. A similar process took place on the other planets, but wasn't carried to completion on all of them. Astronomers have found, by analyzing the light reflected by Jupiter and Saturn, that the atmospheres of these planets still contain large amounts of ammonia and methane.

But while the first atmosphere of the earth was being lost, a new one was rising out of the ground to take its place. Melted rock kept pushing its way up through cracks in the earth's crust, the way it does in volcanoes today. Gases that were dissolved in the melted rock were released at the surface and spread around the globe to form a new atmosphere. This new air consisted largely of water vapor, nitrogen and carbon dioxide.

The Sea

After the earth had cooled some more, its atmosphere became too cool to hold all the water vapor that was in it. The vapor began to condense, and great rains poured out of the sky. The rains continued for years and years. The water flowed down from the high places on the ground and collected where the ground was low, to form the great seas that cover most of the earth's surface. The rain drops dissolved some of the carbon dioxide that was in the air, and carried it down to the sea. So much water vapor and carbon dioxide fell from the air that the nitrogen that remained behind made up most of the air that was left.

The Raw Materials of Life

Water is a universal solvent. When rain drops formed in the air, they washed carbon dioxide and other soluble compounds out of the air, and carried them down to the ground and into the sea. On the ground, streams flowing over the rocks rotted them, wore them down, and washed salts out of them. These salts, too, were carried into the sea. There the salts from the ground and the chemicals from the air mixed and reacted to form new chemical combinations.

Among the chemicals washed into the sea were carbon and nitrogen compounds that served as the raw materials from which living things developed. Some hydrocarbons and ammonia may have been left from the original atmosphere of the earth. But, in addition, there were natural processes going on that were capable of producing them. Some of these were described in Chapter III. Volcanic eruptions were bringing melted rock or lava up to the surface of the ground. At the surface, carbides in the lava reacted with water to form hydrocarbons. The carbides also reacted with nitrogen to form cyanamides. Then the cyanamides reacted with water to form ammonia. There were also nitrides in the lava. These, too, reacted with water to form ammonia. In the upper atmosphere, sunlight was separating hydrogen atoms from the hydroxyl groups to which they were joined in water molecules. Some of these hydrogen atoms combined with nitrogen to form ammonia. In all these ways a supply of hydrocarbons and ammonia was being built up and carried into the sea.

There was another process, too, that was producing organic compounds. There were fast-moving particles called *cosmic rays* crashing into the earth's atmosphere from surrounding space. When they collided with molecules of the air, they struck them with such force that sometimes they pushed small molecules together to make bigger molecules. Laboratory experiments have proved that, when fast-moving particles like these crash through a mixture of water vapor and carbon dioxide, they can produce formaldehyde and other organic compounds.

Water, hydrocarbons, and ammonia are the raw materials out of which amino acids can be made. Two forces which existed in the air in those days are known to be able to make them. The sunlight that streams into the air contains *ultra-violet rays* (the rays that cause sunburn). Ultra-violet rays are hard-hitting rays that carry a large amount of energy. Laboratory experiments have proved that this energy is capable of tearing atoms out of water, hydrocarbons, and ammonia and joining them to form amino acids. Another energy source that can do the same thing is an electrical discharge like *lightning*. The ultra-violet rays of daylight and the lightning discharges of great storms produced amino acid molecules in the air, and the rains washed them into the sea.

Organic Soup

With all these organic chemicals in it, the sea was like a great bowl of organic soup. The organic ingredients of this soup were mixed and combined in its warm waters, and slowly cooked. Out of this natural cookery, more and more complicated molecules grew up. All of the chemical reactions by which molecules are changed, or by which small molecules are joined to form larger ones, took place in the sea. Oxidation-reduction reactions formed aldehydes, ketones, alcohols and organic acids. Condensation reactions joined carbon atoms directly to carbon atoms to form long carbon chains and rings. Polymerizations linked carbon atoms with each other through oxygen or nitrogen bridges.

In living cells these reactions take place with the aid of enzymes that serve as catalysts. You may wonder how they could take place in the sea before these enzymes existed. But catalysts do not *cause* a chemical reaction. They merely *speed it up.* Although the catalysts were not there, the reactions were able to take place anyhow, only much more slowly. And even though the chemical reactions were slow, they were able to produce great changes because they had plenty of time in which to do it. Even the slowest reaction can produce noticeable results in two billion years!

However, even in those early days there were catalysts to help. The waters of the sea were made muddy by particles of clay washed down from the land. Tiny clay particles have the power to adsorb large numbers of molecules on their surfaces. The molecules brought into contact in this way reacted more quickly to form new compounds. In addition, some of the newly formed compounds served as catalysts for particular reactions.

Energy is needed to bind small molecules together into large ones. This energy was supplied by ultraviolet rays, which fell into the sea at that time in greater quantities than they do today. Today there is oxygen in the air, and some of this oxygen is high up in the atmosphere in the form of *ozone.* The ozone serves as a filter that removes most of the ultra-violet rays from the sunlight before it reaches the ground. But before life existed, there was practically no free oxygen in the air. There was no ozone filter, and nearly all the ultra-violet light that entered the atmosphere reached the ground and the sea.

The molecules that were able to combine in the greatest number of ways were the amino acids. By joining together in peptide chains, they built a variety of protein molecules. But protein molecules are large enough to be colloidal particles. As the number of proteins grew, the organic soup in the sea was changed from an ordinary solution to a colloidal solution.

Chemical Mixtures with a Skin

Colloidal particles in a solution, as we saw in Chapter VI, tend to come together or coagulate into lumps large enough to sink. At the same time there are forces that tend to keep the particles apart. The two opposing tendencies sometimes lead to a compromise in which the particles are not scattered, but, on the other hand, do not coagulate. This compromise, also described in Chapter VI, is the formation of colloidal clusters known as coacervates. *The formation of coacervates in the ancient seas opened up a new phase of the chemical evolution that led to the origin of life.* In this new phase some of the characteristics of protoplasm already began to appear.

While colloidal particles were scattered throughout the sea the chemical mixture which they formed had no separate identity apart from the sea. The mixture was not only *in* the sea. The mixture *was* the sea. But the particles that came together to form a coacervate were separated from the rest of the sea. *Each cluster of particles occupied a definite region of space* separated from its surroundings by a thin skin of molecules of water. This was an important step towards the development of protoplasm, which also exists in the form of colloidal clusters separated from their surroundings by a skin.

Clusters that Last

The skin surrounding a coacervate was like a wall separating it from the rest of the world. At the same time, it was like a passageway connecting it with the surrounding world. There was a steady flow of molecules through the skin. Molecules floating in the sea passed through the skin and joined the cluster inside. Others broke away from the cluster and passed out into the sea again. This flow of molecules kept changing the chemical and electrical conditions inside the cluster.

The changes in conditions affected the balance of forces inside the cluster. Some of the changes led to a breakdown of the large molecules in the cluster, and strengthened the forces tending to separate the particles. If the changes were unopposed, then the cluster broke up, and the particles scattered in the sea. No cluster could escape the influence of this tendency to break up. But in some coacervates, while the tendency to break up existed, other changes tended to rebuild the cluster and hold it together. If the process of rebuilding went on at least as fast as the process of breaking down, then the cluster was *stable* and lasted a long time. If the building-up process was faster than the breaking-down process, the coacervate *grew*. The particles of coacervates that broke up were captured sooner or later by coacervates that were stable and grew.

In the development of stable coacervates we see the appearance of two features that were important in the origin of life. One of them was the existence side by side of balanced processes of breaking down and building up. As coacervates evolved, these processes were perfected and organized into a co-ordinated system of chemical changes that later became the metabolism of living cells. A coacervate continued to exist only as long as this balanced system of processes continued to operate. But the system could operate only in the molecules needed for the building-up process continued to flow into the cluster. Stable coacervates needed "food" to maintain themselves and grow on, just as living organisms do. The other feature was the beginning of a process of natural selection. The less stable coacervates broke up and disappeared. The more stable ones persisted and grew. Here we see the first operation of the law of the *survival of the fittest*. The operation of this law led to the continued evolution of the more stable coacervates.

The Faster the Better

Even stable coacervates didn't last forever. But those lasted longest that were most efficient in absorbing and keeping the "food" molecules that surrounded them. The efficiency of a coacervate depended very much on the *speed* with which its chemical processes took place. One in which the building-up processes took place faster would grow faster and get more of the surrounding "food" than a coacervate in which chemical changes took place slowly. In some of the coacervates, molecules that wandered into the cluster served as catalysts for the processes going on inside it. These coacervates had an advantage over their neighbors, grew faster, and lasted longer. Sometimes additional molecules were adsorbed on the surfaces of these catalysts and changed them. If the adsorbed molecules served as inhibitors, they spoiled the action of the catalyst, and the coacervate lost its advantage. If they served as promoters, they improved the action of the catalyst, and gave the coacervate an even greater advantage. As a result, the coacervates with the best catalysts lasted the longest. The survival of the fittest led to a gradual evolution of the catalysts within the coacervates, leading finally to the development of the enzyme systems we find in living things today.

Divide and Conquer

Some of the coacervates, after growing a while, split into two pieces. This had an important effect on their ability to grow. As we saw on page 103, when a particle is broken into pieces its surface is increased. Since the coacervates got their "food" through their surfaces, the more surface a coacervate had, the faster its "food" could flow in. So when a coacervate split into two parts it could grow faster. Those coacervates that had a tendency to split had an advantage over those that didn't, and gradually displaced them.

Enzymes that Copy Themselves

Among the many enzymes that developed, some emerged that had the ability to make copies of particular complex molecules. Most interesting of all were those that could make copies of themselves. Coacervates that contained these self-reproducing catalysts had an advantage over all others. When they split into two, each "daughter" coacervate got a supply of these catalysts. As a result, the new coacervates were like the old ones from which they were formed. Because of this reproduction, that *type* of coacervate continued to exist even after all the original coacervates like it had broken up.

A Struggle for Existence

As millions of years went by, more and more of the organic compounds in the sea water were drawn into coacervates. After a while here was not enough dissolved organic material left to feed all the coacervates that existed. The shortage of "food" doomed many of them to destruction. Those that grew did so at the expense of others. A real "struggle for existence" developed which speeded up the process of natural selection. Only the coacervates with the most stable metabolism, with the most highly developed system of enzymes, and with the most efficient system of reproduction could survive. In this way, the metabolism, the enzyme systems, and the mechanisms of reproduction of the coacervates gradually became more and more efficient. Through the slow accumulation of small changes during hundreds of millions of years, coacervates finally developed that had the complex, delicately balanced, and stable chemical processes that we call life.

The Beginning of Life

These first living organisms were probably even simpler in structure than bacteria or blue-green algae. Like the coacervates from which they developed, they got their food from the organic compounds dissolved in the water that surrounded them. So they were heterotrophic organisms, dependent on ready made food, and unable to make it themselves out of inorganic compounds. Part of their food served as fuel to supply the energy they needed. They burned their fuel by processes of fermentation which were anaerobic—that is, they could take place without the help of free oxygen. Fermentation of sugar is an inefficient way of getting energy, because, as we saw in Chapter V, it releases only a small part of the energy that is stored in the sugar. The more efficient aerobic process of burning that uses free oxygen was impossible at the time because there was no free oxygen around. Besides, the more efficient process was itself the outcome of a long period of evolution. However, even with the most primitive kind of system for releasing energy, the first organisms had to have some way of storing the energy and transferring it to the processes where it was used. So ATP or something like it must have been present in the earliest forms of life.

The Great Food Crisis

As they grew and multiplied, the organisms gobbled up more and more of the organic compounds in the sea. The shortage of food, which had already had an effect in speeding up the evolution of coacervates, became even more serious. Many organisms, unable to get the food that they were used to, starved and died. Some, because of slight changes in their chemistry, were able to make the food they needed from somewhat simpler organic compounds that were still available. But when these compounds, too, disappeared, again some starved, while others managed to survive by developing a way of making these compounds too. A kind of backwards evolution of chemical processes developed. When they could no longer get compound A, some organisms used compound B to make compound A. When they could no longer get compound B, some used compound C to make compound B and then they made compound A. The process of food building developed as a chain of steps. Finally, some organisms appeared that were able to start the chain by using inorganic compounds. These were the first autotrophic organisms, able to make their own food. The organisms that developed this ability saved themselves from starvation. But they saved the heterotrophic organisms too. The food makers lived on the food they made. And the food takers began to live by eating the food makers.

The evolution of autotrophic organisms proceeded in many directions. Some organisms developed the ability to build their food out of carbon dioxide and hydrogen sulphide. They were the ancestors of the sulphur bacteria of today. Others became able to make direct use of ammonia. They were the ancestors of the nitrifying bacteria. Finally one group appeared that could carry out photosynthesis, making glucose out of carbon dioxide and water. They were the ancestors of the green plants of today.

The Effects of Photosynthesis

Photosynthesis led to great changes in the world. In photosynthesis, water molecules are broken up into hydrogen atoms and oxygen atoms. The hydrogen atoms are built into the structure of glucose. But the oxygen of the water is released as a gas into the air. This led to a gradual change in the atmosphere. The amount of oxygen in the air began to increase, even though some of the oxygen was withdrawn from the air by iron in the ground. As a result, one fifth of the air today is oxygen. At the same time, the amount of carbon dioxide in the air began to decrease. In photosynthesis, carbon dioxide is withdrawn from the air and built into the body of plants. Some of this carbon dioxide is returned to the air when plants decay. But some was never returned, because large numbers of plants were buried in the ground to form coal.

The appearance of free oxygen in the air paved the way for the development of aerobic organisms that used oxygen to burn their fuel. Burning with oxygen was better than fermentation because it released more of the energy that was hidden in the fuel. This extra energy made it possible for organisms to grow larger and more complicated. It was the foundation of the evolution of the many-celled animals and plants of today.

Most of the oxygen that accumulated in the air was in the form of O_2, in which each molecule of gas had two atoms of oxygen. But, high in the air, some of the oxygen was converted by the action of sunlight into ozone, in which each molecule has three atoms of oxygen. The ozone layer in the air began to filter out most of the ultra-violet rays in sunlight, and prevented them from reaching the ground. Before this happened, when the full amount of ultra-violet rays in the sunlight reached the ground, it was impossible for living things to exist on land, because the hard-hitting ultra-violet rays would have killed them all. Organisms were able to live in the sea, because the thick layer of sea water above them shielded them from the destructive rays. But when the ozone layer began to stop most of the ultra-violet light before it reached the ground, the open air, too, became safe for living things. *This made it possible for life to come out of the sea for the first time.* Plants and animals began to invade the continents and islands.

Accident and Direction

This is the general picture scientists have formed of how life developed from dead matter. There are big gaps in the picture, but more research will fill in more of the details. There may be errors in it, but they will be corrected by future discoveries.

In this picture, the development of life appears as something that just *happened,* without any design or purpose. It started from the accidental mixing and combining of chemicals in the primitive sea. But the direction of development it took was not all accidental. It was influenced by the natural preferences the chemical elements have for each other. It was built on the basis of carbon's ability to form long-chain compounds. In the later stages of chemical evolution it was also directed by the effects of natural selection. The rule of survival of the fittest guided evolution toward the development of more complicated and more efficient organisms, and finally towards the emergence of intelligent beings.

Could It Happen Again?

A curious thing about the evolution of living things from organic compounds floating in the sea is that once it has happened, it cannot happen again. Organic compounds in the sea have no chance to go through the long process of evolution again, because the living organisms that are in the water eat them before very much happens. The chemical evolution that led to the origin of life could take place only when there were no living things that could interfere with it.

Did It Have to Happen?

Because the origin of life depended in part on the accidental mixing and combining of chemicals, it is natural to wonder whether things had to happen the way they did. Could the chemical changes in the sea have taken another direction, by-passing the steps that led to the origin of life? The answer to this question seems to be, "No." The molecules that were dissolved in the sea were mixed and scattered by the blind forces of nature. They were pushed around by ocean currents and by the dance of the molecules. Whether or not the molecules needed for a particular combination were brought close enough to combine was a matter of chance. But they would combine *only if they were capable of combining. And, given a long enough time,* all possible combinations were bound to appear sooner or later. Then, which of them persisted was no longer purely a matter of chance. The unstable compounds broke up, and the stable compounds remained. Natural selection directed the chemical evolution of organic compounds toward the development of more and more stable combinations. So it is likely that in the main features of its development it had to proceed the way it did. Different accidental circumstances might have led to the development of different forms of life, but, in one form or another, life was bound to come into existence.

Is There Life on Other Worlds?

If life was bound to come into existence under the conditions that existed on the earth, are there other planets where the same conditions existed and the same evolution took place? The answer to this question depends on how planets like the earth may be formed. One theory, as we saw, says that the earth was formed when a passing star came close to the sun and pulled out a large piece of it. On the basis of this theory, other stars might have planets like the earth only if they, too, had a near-collision with a passing star. But the stars are so very far apart that near-collisions of this type must be very rare. If this is really the way the earth was formed, it is very unlikely that other stars have planets on which life could develop.

However, the near-collision theory of the origin of the earth is probably wrong. Planetary systems are not so rare. In fact, the study of the stars has shown that most of the stars are double or even triple stars, revolving around each other the way the planets revolve around the sun. So far it has been possible to explain the existence of double stars only by means of the theory that stars and planets were formed together from whirling clouds of dust and gas in space. But if this theory is true, then it is not only possible, but even likely that there are many other stars that have planets. Let us suppose that only one star in a million has planets. Suppose that, of these, only one in a thousand has a planet like the earth where life could develop. Suppose, too, that among the planets where life could develop, it actually did develop into high forms of life on only one in a thousand. Even then, there are so many stars in the universe that high forms of life would exist on one hundred million different worlds scattered through space! We must admit as a serious possibility the existence of life in other parts of the universe.

Life's Past and Man's Future

Ever since man became aware of the difference between living and dead matter he has been curious about the origin of life. Science, through the co-operation of its different branches, such as biology, chemistry, astronomy, and geochemistry, is beginning to produce facts and theories to satisfy this curiosity. But the study of life's past is more than an attempt to satisfy idle curiosity. It has a great practical importance for the future life of mankind. Understanding how living things came into existence is closely related to understanding how living things work. And the better we know how living things work the better use we can make of them for our own good.

This relationship is shown clearly in some of the research work that is now being carried on. The study of protein structure has at last succeeded in identifying the arrangement of the amino acids in a protein molecule. This will make it easier to learn more about how proteins came into existence in the first place. But it also opens the door to the possibility of making proteins in the laboratory. Think of how important it will be to medicine and industry when we finally learn how to manufacture any enzyme or other protein that we want.

A good deal of work is being done today in the scientific study of viruses, those tiny particles of protein and DNA that seem to be halfway between dead and living matter. So much has been learned about viruses that in 1956 chemists even succeeded in making one. Further study of viruses will cast more light on the way DNA came to play its important part in living cells. But it also opens the door to the conquest of the diseases that viruses cause.

The study of photosynthesis has succeeded in identifying the steps by which it takes place. This new information will lead to a better understanding of how autotrophic organisms originally developed. But it also opens the door to the possibility of carrying out photosynthesis without the help of living cells. Once we know the full secret of photosynthesis, we won't have to rely on green plants to make our food for us. We shall be able to make it ourselves out of air and water in the factory. Unlimited factory production of food would banish hunger forever.

The study of cancer is an important area of medical research. In 1956 one scientist advanced the theory that normal cells become cancer cells when they lose the ability to carry out respiration, the aerobic form of oxidation, and must fall back on the ancient inefficient method of fermentation. Further study of this interesting question may cast more light on the way respiration originally developed. But it may also lead to a cure for cancer and to the lengthening of life.

The study of life's past and the protection of man's future continue to go hand in hand.

Index

Accident, 121-2
Acetaldehyde, 70-6, 90
Acetaldol, 74
Acetamide, 71-2
Acetic acid, 72-4, 75-8, 93-4
Acetone, 70-1
Acetylene, 59, 66-7, 73
Acid, 56, 58, 71, 73, 75, 79, 80, 82, 84, 91, 113
Actomyosin, 94-5
Addition, 72, 74
Adenine, 86, 88, 99
Adenosine diphosphate, 89
Adenosine triphosphate, 88
ADP, 89, 93-4
Adsorption, 104-5, 107, 114, 116-7
Aerobic, 34-5, 94-5, 118, 120, 124
Age of Earth, 109
Alanine, 81-2
Albumin, 107
Alcohol, 341, 68-70, 72-4, 75, 78-9, 84, 90, 113
Aldehydes, 70-2, 78-9, 86, 113
Aldose, 79
Algae, 24
Aluminum, 48, 61, 111
Amide, 72, 78
Amines, 70, 78, 82, 84
Amino acids, 31, 37, 39, 60, 75, 81-4, 87, 94-8, 103, 113-4, 124
Amino group, 70, 82-4, 97, 105
Ammonia, 38-40, 45, 57, 60-1, 70, 76, 111-3, 119
Amphibians, 24
Anabolism, 27
Anaerobic, 34-5, 93-4, 118
Angstrom, 103
Animals, 9, 24, 26, 36
Antibodies, 96
Argon, 47-9
Aristotle, 13
Atmosphere, 111, 119-20
Atoms, 45-51, 101
ATP, 88-90, 92-4, 96, 98, 119
Autotrophic, 31-3, 35, 98, 119, 123

Backbones, 15, 24
Bacteria, 32, 36, 38, 39, 44, 84, 109, 118-9
Base, 57-9
Benzene, 64
Beryllium, 48-9, 62
Birds, 20, 24, 26
Blue-green algae, 39, 44, 109, 118
Boron, 48-9, 62-3
Breath, 9-11, 34
Bushmen, Legend of, 10
Butane, 67-9
Butter, 29
Butyl alcohol, 69

Calcium, 60-1, 66, 111
Cancer, 124
Canine, 15
Carbides, 59, 61, 66, 112
Carbon, 30-1, 34-5, 45, 48-9, 54, 62-86, 112
Carbonate, 56, 60
Carbon cycle, 35-7
Carbon dioxide, 31, 34-7, 45, 54-5, 58-9, 72, 87, 89-91, 92-3, 111-3, 119-20
Carbonic acid, 56
Carbon monoxide, 54
Carbonyl group, 71-2, 78-9, 85
Carboxyl group, 71-2, 79-85, 97, 105
Carboxylic acid, 71-2, 78, 80-1
Catabolism, 27
Catalysts, 61, 66, 87, 90, 114, 116-7
Cause and effect, 11
Cell, 18-9, 24-28, 32-4, 40-4, 61, 107
Cellulose, 79, 92
Cell wall, 40
Chain compounds, 63, 67-8, 74, 78, 83, 95, 98, 113
Chemical bonds, 45, 50-1, 53
Chemical energy, 26
Chloride, 50, 56-8
Chlorine, 48-50, 56
Chlorophyll, 31-2, 40, 45, 85
Chloroplast, 40-1, 89

125

126

127